Conceiving Contemporary Parenthood

Imagining, Achieving and Accounting for Parenthood in New Family Forms

Edited by
Zeynep B. Gürtin and Charlotte Faircloth

Routledge
Taylor & Francis Group
LONDON AND NEW YORK

First published 2021
by Routledge
2 Park Square, Milton Park, Abingdon, Oxon OX14 4RN

and by Routledge
52 Vanderbilt Avenue, New York, NY 10017

Routledge is an imprint of the Taylor & Francis Group, an informa business

British Library Cataloguing in Publication Data
A catalogue record for this book is available from the British Library

ISBN 13: 978-0-367-68743-4

Typeset in MinionPro
by Newgen Publishing UK

Publisher's Note
The publisher accepts responsibility for any inconsistencies that may have arisen during the conversion of this book from journal articles to book chapters, namely the inclusion of journal terminology.

Contents

Citation Information

The chapters in this book were originally published in *Anthropology & Medicine*, volume 25, issue 3 (December 2018). When citing this material, please use the original page numbering for each article, as follows:

Introduction

Conceiving contemporary parenthood: imagining, achieving and accounting for parenthood in new family forms
Zeynep B. Gürtin and Charlotte Faircloth
Anthropology & Medicine, volume 25, issue 3 (December 2018), pp. 243–248

Chapter 1

Being a 'good' parent: single women reflecting upon 'selfishness' and 'risk' when pursuing motherhood through sperm donation
Susanna Graham
Anthropology & Medicine, volume 25, issue 3 (December 2018), pp. 249–264

Chapter 2

The role of normative ideologies of motherhood in intended mothers' experiences of egg donation in Canada
Kathleen Hammond
Anthropology & Medicine, volume 25, issue 3 (December 2018), pp. 265–279

Chapter 3

Accounting for the money-made parenthood of transnational surrogacy
Ingvill Stuvøy
Anthropology & Medicine, volume 25, issue 3 (December 2018), pp. 280–295

Chapter 4

Surrogate non-motherhood: Israeli and US surrogates speak about kinship and parenthood
Elly Teman and Zsuzsa Berend
Anthropology & Medicine, volume 25, issue 3 (December 2018), pp. 296–310

Chapter 5
Ideals, negotiations and gender roles in gay and lesbian co-parenting arrangements
Cathy Herbrand
Anthropology & Medicine, volume 25, issue 3 (December 2018), pp. 311–328

Chapter 6
Conceptions of transgender parenthood in fertility care and family planning in Sweden: from reproductive rights to concrete practices
Jenny Gunnarsson Payne and Theo Erbenius
Anthropology & Medicine, volume 25, issue 3 (December 2018), pp. 329–343

Chapter 7
Commentary
Marcia C. Inhorn
Anthropology & Medicine, volume 25, issue 3 (December 2018), pp. 344–348

Notes on Contributors

Zsuzsa Berend, Department of Sociology, University of California Los Angeles, CA, USA.

Theo Erbenius, Independent Researcher, Sweden.

Charlotte Faircloth, Social Research Institute, University College London, UK.

Susanna Graham, Centre for Family Research, University of Cambridge, UK.

Jenny Gunnarsson Payne, Department of Historical and Contemporary Studies, Södertörn University, Sweden.

Zeynep B. Gürtin, Institute for Women's Health, University College London, UK.

Kathleen Hammond, Faculty of Law, Ryerson University, Toronto, Canada.

Cathy Herbrand, Centre for Reproduction Research, De Montfort University, Leicester, UK.

Marcia C. Inhorn, Yale University, New Haven, CT, USA.

Ingvill Stuvøy, Department of Sociology and Political Science, Norwegian University of Science and Technology, Trondheim, Norway.

Elly Teman, Department of Behavioural Sciences, Ruppin Academic Centre, Emek Hefer, Israel.

INTRODUCTION

Zeynep B. Gürtin and Charlotte Faircloth

Introduction

This year, as the world celebrates the 40th anniversary of the first IVF birth – of Louise Brown, born in Oldham, UK in 1978 – we find ourselves in a reproductive milieu that has been dramatically transformed by the introduction, global expansion and endless development of reproductive technologies. Initially developed as a means to aid the reproduction of married heterosexual couples facing medical infertility (specifically, blocked fallopian tubes), reproductive technologies have since multiplied and diversified, to include gamete donation, gestational surrogacy, fertility preservation and myriad genetic technologies. Correspondingly, there has been a proliferation in the types of family forms made possible by these technologies and an even greater array of moral, ethical and legal responses that have sought to regulate their use. This special issue is dedicated to exploring the experiences of men and women involved in creating these new family forms, with a particular emphasis on how they imagine, achieve and account for parenthood, both at the point of conception, as well as in the practice of daily life. In line with the aims of Anthropology & Medicine, each of the articles focus on the interface of culture and medicine, in particular the interface of parenting culture and reproductive technologies.

Today, in addition to IVF, reproductive technologies are increasingly being used by single women and same-sex couples, as well as by men and women in heterosexual relationships who need the 'third party assistance' of egg donors, sperm donors or surrogates to conceive the babies they imagine and desire. Of course, these forms of reproductive assistance are not permitted or possible in every jurisdiction, but intending parents can increasingly resort to cross-border reproductive care to access treatments that are prohibited in their home countries, or are not readily available due to resource shortages (Gürtin and Inhorn 2011). Although, of course, it is crucial to note that the global landscape of reproduction remains troublingly uneven, particularly along economic lines (Franklin 2011), we are nevertheless living in a context where reproductive technologies enable a variety of new family arrangements, including solo parenthood 'by choice', same-sex parenting, co-parenting, donor-conceived families and families created following fertility preservation.

With a proliferation of reproductive possibilities and arrangements, it is fitting that over the past two decades a growing body of research, particularly across sociology and anthropology, has sought to better understand the cultural impact of reproductive technologies, at both the local and the global level (see, e.g. Inhorn and Birenbaum-Carmeli 2008). This special issue is a contribution to this growing stream of *Reproductive Studies* scholarship. At the same time, it is a contribution to an emergent body of work known as *Parenting Culture Studies* and a part of our recent efforts to create fertile connections across these two fields (Faircloth and Gürtin 2017a). While the former concentrates on the making of babies, the latter focuses on the ways in which they are raised; activities that are clearly interlinked in many complex ways, but which have often been dealt with separately in scholarship. While *Reproductive Studies* has often focused on intending parents' experiences of reproductive technologies and the co-constructed relationships between science and society, *Parenting Culture Studies* has examined the ways in which men and women parent and the changing ideologies that underpin their actions. In particular, *Parenting Culture Studies* has called attention to the 'intensification' of parenting, specifically in Anglophone contexts (but in a globalised world, also beyond), during precisely the same period as the 'explosion' of technologies of assisted reproduction. Far from being common-sense, 'parenting' is now a much more concerted activity, requiring that parents are fully informed and 'ready' before embarking on this important life stage, fuelling a multi-million pound industry of expertise and advice (Hays 1996; Lee, Bristow and Faircloth 2014). Certainly, the way raising children is talked about in these contexts has shifted dramatically in the space of a generation.

Elsewhere (Faircloth and Gürtin 2017a), we provide a more detailed overview of these two bodies of scholarship across the disciplines of anthropology, sociology and social work, and have drawn out four interlinked themes – Reflexivity, Gender, Expertise and Stratification – to promote a more holistic approach to understanding how children are conceived and cared for within the current reproductive landscape. In another special issue, with the contribution of colleagues from across these fields, we have thought specifically about the 'ways in which an increasingly globalized "parenting culture" creates an appetite – explored in different national contexts as well as transnationally – for reproductive technologies, as well as the ways in which these same technologies for "making parents" shape that very culture of "doing parenting". (see Faircloth and Gürtin 2017b). The articles brought together here continue and expand upon that conversation, this time focusing specifically on how intentions, expectations and the burgeoning technologies of reproductive 'assistance' contribute to changing and enduring conceptions of parenthood – in particular of *good parenthood* – in the 21st Century.

Each of the articles presented in this special issue explores a 'new family form' – including solo motherhood using donor sperm; conception with donor eggs; surrogacy; co-parenting arrangements between gay men and lesbian women; and trans parenthood. They examine how, in these variously non-normative scenarios, women and men activate their intentions, reflect on their expectations, account for their choices and narrate their journeys to parenthood (or non-parenthood). As the articles discuss, these families have not only been enabled by technological possibilities and global commercial developments, but also by changing social mores around what makes a good parent, and thus who can make claims and gain access to parenthood, though tensions often continue to exist when there remains a gap between normative reproduction or parenthood and an individual's own experience. The men and women in these arrangements must first conceive of themselves as (good) parents, then advocate for their reproductive rights and account for their parenting decisions and choices.

Situated in a variety of ethnographic settings, including the UK, Canada, Norway, the US, Israel, Belgium and Sweden, as well tracing the cross-border movements of intending parents to different global locations in search of reproductive assistance, the articles highlight with clarity both the expanding opportunities and the increasing demands and challenges that women and men face in their 'quests for conception' (Inhorn 1994). Often, the contemporary journeys to parenthood described in these articles present complex entanglements between culture, commerce and medicine, with ontological choreographies (Thompson 2005) involving not only seeking biomedical assistance, but also economic exchanges, international travel, relational agreements, emotional discussions, personal deliberations and the negotiation of ideological claims to good parenthood.

Whether as single mothers, non-genetically related parents, trans persons, repro travellers, or men and women in family arrangements with non-conjugal partners, the actions of the protagonists in these new family forms simultaneously replicate and revolutionise normative reproduction and parenting ideologies. In *Conceiving the New World Order*, Ginsberg and Rapp write that 'cultures are produced (or contested) as people imagine and enable the creation of the next generation [...] regardless of its popular associations with notions of continuity, reproduction also provides a terrain for imaging new cultural futures and transformations' (Ginsberg and Rapp 1995, 1–2). As such, the collection addresses fundamental issues, not only around the changing possibilities of reproduction, but also around the changing normative discourses around (good) parenting, family formation and gender.

Overview of articles

Being a 'good' parent: single women reflecting upon 'selfishness' and 'risk' when pursuing motherhood through sperm donation – ***Susanna Graham***

The first article in our collection, by Susanna Graham, focuses on single heterosexual women's experiences of thinking about or pursuing solo motherhood through sperm donation. Based on interviews with 23 women in the UK, Graham unpacks the ways in which the women felt solo motherhood to be a departure from their imagined lives of having a child within the context of a stable relationship and brings to light their negotiations with the risks of becoming a mother in this manner, 'with no rule book to follow'. It is particularly interesting that, as Graham notes, 'Although the participants feared others would judge their decision to become a solo mother, many were aware that this judgement was largely coming from themselves'.

The role of normative ideologies of motherhood in intended mothers' experiences of egg donation in Canada – ***Kathleen Hammond***

In the second article, Kathleen Hammond turns our attention to a different group of women battling with the gap between their imagined our assumed routes to parenthood and their lived realities, this time drawing on interviews with 18 Canadian women who used donor eggs to conceive. Examining in particular how these (intending) mothers' talk about and relate to their donors, Hammond proposes a 'relational spectrum' along which mothers' relationships with their egg donors can be conceptualized. 'Intended mothers' reactions to the news that they would have to use donor eggs are replete with references to a normative ideology of motherhood', Hammond tells us, but she goes on to show how these women, in the long run, 'are likely to alter cultural attitudes and conceptions of 'normality' when it comes to reproduction and motherhood'.

Accounting for the money-made parenthood of transnational surrogacy – ***Ingvill Stuvøy***

The third and fourth article in this issue focus on the practice of surrogacy, but from opposing lenses. Ingvill Stuvoy, based on her research with 21 Norwegian heterosexual and same-sex couples and single men and women seeking surrogacy abroad, asks in particular how money is accounted for in specific ways to confirm parenthood in surrogacy arrangements. Focusing our attention on two of the most contentious aspects of assisted reproduction in the 21st Century, Stuvoy adds much-needed nuance and the voices of intending parents to debates over commercialization and reproductive travel. Using an in-depth exploration of three case studies, Stuvoy shows the ways in which men and women make sense of both money and parenthood in particular ways, depending on the specificity of their surrogacy journeys, showing that 'different transnational [surrogacy] arrangements make different kinds of monetized parenthood'.

Surrogate non-motherhood: Israeli and US surrogates speak about kinship and parenthood – ***Elly Teman*** *and* ***Zsuzsa Berend***,

Shifting our perspective from the intending parents pursuing surrogacy to the surrogates themselves, Elly Teman and Zsuzsa Berend offer us an illuminating insight into how non-parenthood is also negotiated. Based on Teman's research with Israeli and Zsuzsa's with US-based surrogates, the authors create not only a comparison of the thoughts, feelings and accounts of surrogates in these two global hubs of surrogacy, but also between the assumed and experienced realities of being a surrogate. The article challenges commonly held myths about surrogacy, showing that surrogates create 'non-motherhood' while simultaneously engaging in 'a labour of love', delineating clear boundaries between their families and those of the intending parents. As the authors conclude, 'The two studies together show that both Israeli and US surrogates consider surrogacy as a morally meaningful undertaking that creates families; surrogates "nurture parents", not just "their babies"'.

Ideals, negotiations and gender roles in gay and lesbian co-parenting arrangements – ***Cathy Herbrand***

The fifth article in the special issue explores an understudied 'new family form' – co-parenting, a set-up typically involving individuals who are not conjugal partners committing to conceiving and raising a child together. Herbrand's research, based on nine case studies of co-parenting arrangements between lesbian and gay couples in Belgium, at once shows the revolutionary and liberatory potential of such set-ups, but also points to the persistence of normative ideologies, in particular around the importance of intensive embodied motherhood. As Herbrand notes, 'while co-parenting offers the potential for parents to reconcile contradictory social expectations and provide them with opportunities to create family practices that suit them, these are restrained by existing gender norms'. These set ups therefore have the potential to challenge and reinscribe contemporary ideologies of 'good parenting'.

Conceptions of transgender parenthood in fertility care and family planning in Sweden: From reproductive rights to concrete practices – ***Jenny Gunnarsson Payne** and **Theo Erbenius***

The final article brings together several of the themes of the special issue, in highlighting Swedish transgender patients' politically mobilization for access to fertility treatments. Based on interviews with medical staff, LGBTQ-advocates and complementing qualitative data, Gunnarsson Payne and Erbenius take a historical perspective on Sweden's treatment of transgender patients vis-à-vis fertility preservation in the process of transitioning, and explore the question of the 'right' to parenthood, tracing the emergence of this in conjunction with wider historical shifts in parenting culture and the availability of assisted reproductive technologies. As they note, the translation of new reproductive rights for trans men and women into the practical context of a fertility clinic is not always a seamless process, but the growth of this new patient group has nevertheless meant that 'the cultural imagination of who can become a parent and what a family might look like is becoming further destabilised'. They end with a call for renewed political debate around the forms of stratification and inequalities in access to reproductive support, in which transgender status is often a neglected category.

In sum, the articles presented here engage with the new technological, social and commercial developments in the field of reproductive medicine, and reflect in particular on the many complex ways in which these intersect with both enduring and changing ideologies of 'good' parenting. Strongly grounded in anthropological and social theory and methods, the research presented uses ethnography and interviewing to develop deep insights into the experiences not only of parents and intending parents in new family forms, but also reflect the perspectives of surrogates, medical professionals and activists. We are honoured that the articles are followed by an insightful Commentary from Professor Marcia Inhorn, whose pioneering work in the field of reproductive studies has been both an inspiration and an anchor for subsequent research in the field. Read at this critical moment, when the emergence of new reproductive possibilities is met with the simultaneous (re)emergence of conservative challenges to reproductive rights and liberties, we believe that the qualitative details provided in these articles are both timely and political. We hope that this special issue will be of interest to an international readership of medical anthropologists, medical sociologists, social anthropologists and scholars of gender, reproduction, parenting and family life. We hope to hear from those with an interest in pursuing similar lines of enquiry and scholarship in the future.

References

Faircloth, C., and Z. Gürtin. 2017a. "Fertile Connections? Thinking across Assisted Reproductive Technologies and Parenting Culture Studies." *Sociology* 52: 1–18. doi: 10.1177/0038038517696219

Faircloth, C., and Z. Gürtin. 2017b. "Introduction – Making Parents: Reproductive Technologies and Parenting Culture across Borders." *Sociological Research Online* 22 (2): 6. doi: 10.5153/sro.4310

Franklin, S. 2011. "Not a Flat World: The Future of Cross-Border Reproductive Care." *Reproductive Biomedicine Online* 23 (7): 814–816.

Ginsberg, F., and R. Rapp. 1995. *Conceiving the New World Order: The Global Politics of Reproduction*. Berkeley: University of California Press.

Gürtin, Z. B., and M. C. Inhorn. 2011. "Introduction: Travelling for Conception and the Global Assisted Reproduction Market." *Reproductive Biomedicine Online* 23 (5): 535–537.

Hays, S. 1996. *The Cultural Contradictions of Motherhood*. New Haven, CT: Yale University Press.

Inhorn, M. C. 1994. *Quest for Conception: Gender, Infertility, and Egyptian Medical Traditions*. Philadelphia: University of Pennsylvania Press

Inhorn, M. C., and D. Birenbaum-Carmeli. 2008. "Assisted Reproductive Technologies and Culture Change." *Annual Review of Anthropology* 37 (1): 177–196.

Lee, E., J. Bristow, and C. Faircloth. 2014. *Parenting Culture Studies*. Basingstoke: Palgrave Macmillan.

Thompson, C. 2005. *Making Parents: The Ontological Choreography of Reproductive Technologies*. Cambridge, MA: MIT Press.

Being a 'good' parent: single women reflecting upon 'selfishness' and 'risk' when pursuing motherhood through sperm donation

Susanna Graham

ABSTRACT
Drawing upon the narratives of 23 single heterosexual women in the UK thinking about and pursuing motherhood through sperm donation, this paper explores how solo motherhood can be construed as a 'risk' to the identity of a 'good' mother. It shows how, for these women, solo motherhood was a departure from an imagined life of having a child within the context of a stable relationship and was a prospect viewed with much ambivalence and uncertainty. *Choosing* to become a single mother challenged their conceptualisation of a 'good' mother, someone who puts their child's interests above their own. By examining the flexible and creative ways these women navigate the field of reproductive technologies in the pursuit of motherhood, this paper explores how the women engaged in a process of 'damage limitation', becoming 'moral pioneers' of family and relatedness, as they realigned their ideals versus realities of family building and tried to make solo motherhood as 'good as it can be' for their future child.

Introduction

'Solo mothers', also referred to as 'single mothers by choice' and 'choice mothers', are women who actively become mothers without the involvement of a partner. They can achieve motherhood in a number of ways but many do so through attending fertility clinics for treatment with donor sperm (Jadva et al., 2009). This paper explores the narratives of twenty-three single, heterosexual women in the UK thinking about and becoming solo mothers through sperm donation. In doing so it explores how, for these women, solo motherhood posed a 'risk' to the identity of a 'good' mother. As a departure from an imagined life of having a child within the context of a relationship, solo motherhood was a prospect viewed with much ambivalence and uncertainty. Was solo motherhood 'fair' on a child or were they being 'selfish', putting their need for an 'own child' above their child's need for a father? Could they actively *choose* single motherhood yet still construct themselves as a 'good mother', someone who puts their child's needs above their own? In exploring these women's accounts of their thoughts and feelings about becoming a solo mother we see how they normalise and justify their decision through deconstructing the nuclear family ideal. However, solo motherhood was seen to not only challenge the structure of the nuclear family but also its ideology, in particular the perceived need to know one's 'genetic origins'. By examining the flexible and creative ways the participants navigate the field of reproductive technologies in the pursuit of motherhood, this paper explores how the women engage in a process of 'damage limitation', becoming 'moral pioneers' (Rapp 1999) of family and relatedness, as they realign their ideals with the realities of family building and try to make solo motherhood as 'good as it can be' for their future child.

Methodology

The data in this chapter was collected as part of PhD research exploring the decision-making and experiences of single women in the UK thinking about and becoming mothers through sperm donation. A convenience sample of twenty-three participants was recruited from the Donor Conception Network (DCN),[1] Fertility Friends,[2] and the London Women's Clinic.[3] All participants were pursuing motherhood by accessing fertility clinics for treatment with sperm from an unknown donor. All identified as heterosexual and were aged between 33 and 46 years. Twenty-one participants identified as white, one participant as black (African) and one as Asian (Chinese). All were highly educated, in professional employment and financially independent at the time of seeking treatment.

Multiple interviews were held with the women over a fifteen month period between 2010 and 2012. Two participants were interviewed once, eight participants on two occasions, nine participants on three occasions and four participants were interviewed four times. The number of interviews held depended upon the participants' own thinking and treatment trajectory, as well as their willingness and time available for interviews. Three of the twenty-three participants did not start fertility treatment during the study period, two decided to stop treatment and six were continuing or considering their options for further treatment when the study period ended. Due to the long and often complex nature of their treatment journey, participants undertook various fertility treatments at various locations, undergoing a combination of intrauterine insemination (IUI), *in vitro* fertilisation (IVF), or in three cases IVF with donor eggs. Eighteen participants had some or all of their treatment in the UK and six had some or all of their treatment abroad. One participant had her fertility treatment funded by the NHS. However, all other participants were private patients at NHS or private fertility clinics, fully funding their own fertility investigations and treatment.

Solo motherhood – parenting culture and ARTs

In pursuing motherhood through sperm donation solo motherhood brings into question not only the structure of the family form created, i.e. a single parent family, but also how that family is made. The use of Assisted Reproductive Technologies (ARTs) opens the decision to become a single parent to scrutiny, not only by women contemplating this family form, but also policy makers, 'experts' working in the health and fertility industries, and external others. When single women pursue motherhood through sperm donation a single parent family is not formed through circumstance, i.e. the breakdown of the nuclear family, but actively created from the outset (Graham and Braverman 2012).

Risk and accountability in good parenting

Faircloth and Gürtin (2017), in their paper developing a dialogue between the scholarship of 'parenting culture studies' and the social and anthropological research concerned with the social, cultural and ethical aspects of the globalisation of ARTs, argue that a culture of 'anxious reproduction' has been an unintended and paradoxical consequence of the greater availability of reproductive technologies and an increasingly 'intensive' parenting culture. Drawing upon this theme, this paper will show how, when considering motherhood through sperm donation, solo mothers can be viewed as 'pre-conception' parents (Faircloth and Gürtin 2017): in seeking fertility treatment they are expected to account for and embody an intensive commitment to parenting before even becoming parents.

Raising children has become framed as an increasingly complex task: far from simply ensuring the transition to adulthood, today's parents are expected to do much more to protect and optimise the development of their children (Lee et al. 2014). 'Parenting' is therefore no longer considered 'just what parents do'. Instead, what parents feed their children, when they put them to bed, what they read to them, how they discipline them and how they play with them have all become contested and politicised questions (Lee, Macvarish, and Bristow 2010). Indeed, such activities have become the subject of intense debates about the effects parental activities may have on the next generation (Bristow, 2009). Lee, Macvarish, and Bristow (2010) highlight how the expansion of the childrearing task has taken place in tandem with the development of the notion that children are at 'risk'. They argue that while some groups of children are deemed more 'at risk' than others, it is notable how this concept is now applied to all children: parenting has become associated with 'risk-consciousness' (Beck, 1992; Giddens 1999) and parents must act as 'risk managers' to avoid their child's development being compromised. With research consistently showing children raised in single parent families experiencing more cognitive, emotional, and behavioural problems than children living with married parents (Amato 2005; Coleman and Glenn 2009; Dunn et al. 1998; Pryor and Rodgers 2001), it may be assumed that children of solo mothers will be considered particularly 'at risk'. Murphy (2003) argues that in a neo-liberal era, with its emphasis on self-management, 'good' mothers are reflexive, informed consumers able to 'account' for their parenting strategies. As 'pre-conception parents' solo mothers must therefore 'account' for their decision to become a solo mother, be reflexive of the 'risks' that this family form may pose, and align their route to parenthood, and their expected parenting strategies, with that of being a 'good' mother.

Buttney (1993) states that to be accountable to others arises from the condition that persons can be held responsible or answerable for their actions. He describes three types of 'failure' one can be held accountable for: instances of defective skill or ability; insufficient care or effort; and improper intentions. Buttney argues that the most serious kind of 'failure' involves 'improper intentions': 'failures' due to a lack of ability reflect the actor's competence, but failures due to improper intentions reflect the actor's moral character (1993, 4). Indeed. as this paper will explore, it is the *intention* of solo mothers to form a single parent family that is called into question by both the women who are themselves embarking upon this family form, as well as external others, and it is this intention for which they must be accountable. What counts as a 'good' or 'acceptable' account depends, of course, upon our socio-cultural assumptions, norms and ideologies of right, practical, or permissible conduct. This paper will explore the socio-cultural assumptions, norms and ideologies utilised by a certain group of single women in the UK to account for their decision to become a solo mother.

Kinship, relatedness and moral pioneering

As 'pre-conception parents' (Faircloth and Gürtin 2017), solo mothers must also negotiate, and account for, their chosen method of conception, namely the incorporation of donated sperm into their quest for motherhood. How will a child feel about being donor conceived? What sort of sperm donor should they choose? Is donor anonymity beneficial or detrimental to a child? Will a child be interested in their donor and what sort of information would be useful to know about him? In exploring the participants' narratives of their decision to pursue motherhood through sperm donation, this paper therefore also engages with, and adds to, the literature on 'reproductive donation' (Richards, Pennings, and Appleby 2012), kinship and relatedness.

A growing body of anthropological work on ARTs has shown how developments in infertility treatment, as well as the use of donated gametes, have raised new issues and concerns relating to notions of kinship and relatedness (Franklin and McKinnon 2001; Strathern 1992). Edwards and Strathern argue that Euro-American kinship can be conceptualised as a 'division and combination of social and biological facts' (2000, 159). In contrast, Mohr describes how 'relatedness' might be understood as 'any combination of elements (e.g. biological material, symbolism, affective relations, everyday practices) with the capacity to generate meanings and experiences of being related' (2015, 472). Transference of biogenetic substance from a gamete donor to a donor-conceived child may therefore be culturally understood to constitute bonds of relatedness (Carsten 2000). Equally, if relatedness is constructed as everyday practices of 'doing family' (Morgan 2011) then a gamete donor may be construed as having little, or no, significance in a child's life. Indeed, studies have shown how the cultural significance of a gamete donor may vary depending upon the context in which donor gametes are used. Drawing upon anxieties regarding infertility and norms of fatherhood placing importance on genetic connections, the sperm donor may be perceived as challenging the role and place of the social father in a heterosexual couple conceiving with donor sperm (see, for example, Kirkman 2004; Inhorn 2006). Single women and lesbian couples also face questions regarding the significance of a sperm donor, but here it is the *absence* of a social father that questions the sperm donor's significance (Graham 2014; Hertz 2006; Mamo 2005; Nordqvist 2010). Regulations regarding donor anonymity also play a role in the construction of a donor's meaning (Graham, Mohr, and Bourne 20016). Mohr (2015) found that Danish sperm donors make sense of connections to donor-conceived individuals as a particular kind of relatedness that cannot be reduced to either contractual or kinship relations. Almeling (2011), however, describes how sperm donors in the United States think of themselves as 'fathers' to donor-conceived children. She argues that in a culture that equates male genetics with parenthood, and an institutional logic that prioritises identity-release programmes, sperm donors cannot help but see themselves as fathers.

In this paper I suggest Rapp's (1999) concept of 'moral pioneers' is a useful framework within which to explore the narratives of the single women in this study, as they grapple with the meaning of incorporating donor sperm into their journey to motherhood. In her ethnography '*Testing Women, Testing the Fetus: The Social Impact of Amniocentesis in America*', Rapp describes how she 'came to think of the women who submitted to the discipline of a new reproductive technology in order to reap its biomedical benefits as moral pioneers. At once conscripts to techno-scientific regimes of quality control and normalisation, and explorers of the ethical territory its presence produces, contemporary pregnant women have become our moral philosophers of the private' (1999: 306). Just as women undergoing amniocentesis are forced to judge the quality of their own foetuses, making concrete and embodied decisions about standards for entry into the human community, women considering sperm donation as a means to motherhood are forced to consider and negotiate the meaning and implications of donor conception for both themselves and the child conceived. With no rule book to follow with regard to how to choose a sperm donor and no clear evidence regarding what, if anything, their child may be interested to know about their donor, these women can be seen as 'moral pioneers', 'philosophers of the private', entering unchartered waters, whilst they navigate the world of ARTs and gamete donation and try to align their chosen route to motherhood with their aim of being a 'good parent'.

Solo motherhood – a selfish choice

Studies have found that solo motherhood is only considered when women are faced with limited prospects of finding a partner with whom to have a child in the time frame their increasing age and decreasing fertility allows (Mannis 1999; Bock 2000; Murray and Golombok 2005; Hertz 2006; Jadva *et al.* 2009). The narratives of the twenty-three women taking part in this study further elucidate the ambivalence many feel in pursuing this 'choice'. Single motherhood was never the plan but an option that had to be incorporated into their life trajectory if they were to pursue motherhood (Graham, 2012):

> Maybe there are single women out there who are completely happy being single and never want to be in a relationship but I think underneath it this is always either Plan B or C, or even Plan Z for most people. For me this is definitely Plan Z. (Catherine[4]., 36 years old)

All participants expressed a deep-seated desire to become mothers and all had assumed they would one day have a child of their 'own' (Graham, 2012). However, the expectation of motherhood had been premised upon being in a long-term, stable relationship: they assumed they would be mothers within the context of a nuclear family. Nineteen of the participants had been in long-term relationships they thought might progress to children. The four participants who hadn't been in long-term relationships kept 'moving the goal posts', thinking they would still 'meet somebody, settle down, have a family'. Indeed the decoupling of motherhood from the context of a relationship only seemed possible when the participants realised that 'even if [they] met someone tomorrow it would be too late': the contemplation of solo motherhood was triggered by a feeling of 'time running out'. Although social age was important and many of the participants were concerned about being an 'older mother', 'time running out' was predominantly articulated in terms of reproductive ageing. Thus it was the participants' 'biological clock' that was driving their contemplation of solo motherhood. Three of the participants went as far as discussing this sense of time running out through the metaphor of 'a gun against my head'. Similarly, Catherine, who discovered she had 'low fertility' after undertaking a 'fertility M.O.T.'

stated that 'it wasn't until I found out that I *had* to' (my emphasis), that she started considering solo motherhood.

The nuclear family they had imagined for themselves remained at the forefront of the participants' minds, when considering if, and how, they should pursue solo motherhood. However, it wasn't solo motherhood's departure from the nuclear family per se that troubled the participants but the fact that in pursuing motherhood through sperm donation they were *knowingly* departing from this idealised family at the outset. Despite their ambivalence regarding the amount of 'choice' they had in pursuing solo motherhood, all participants were aware they were making a conscious decision to become a single mother, one they spent months, or in some cases years, contemplating. *Knowingly* forming a family that departed from the nuclear family was the aspect of solo motherhood the participants believed ran the possibility of them being constructed as 'poor mothers', 'selfish' for putting their own interests above those of a child:

> Well we are criticised for choosing to bring children into the world without a father basically. And there is still a huge amount of the population that believes that the ideal thing for any child is a mother and a father living together in a family and that's our cultural norm in the UK, that's what you do. So I think ... they see it very much as a selfish choice. (Emma, 40 years old)

Although the participants feared others would judge their decision to become a solo mother, many were aware that this judgment was largely coming from themselves. Aligning with Murphy's (1999) assertion, in the context of infant feeding, that a 'good' mother prioritises her child's needs above her own but a 'poor' mother places her own needs, preferences and conveniences before her baby's welfare, the participants struggled to reconcile solo motherhood with the idealised image of the 'selfless' mother:

> So for me, going down this road, yeah the biggest struggle is that I know what I want but does the child I'm bringing into the world want this? Am I doing right by the child or am I just being selfish? (Vicky, 37 years old)

Sally, a 33-year old who was pursuing initial investigations into solo motherhood when I first met her, described the 'never ending circles' of turmoil she felt in contemplating fertility treatment with donor sperm. She believed that prioritising a child's needs is what 'being a mother is all about' and questioned whether in pursuing solo motherhood she would be doing this. Would she be 'putting any child's feelings and needs above [her] own' or merely following her own desire for motherhood at the expense of her child? For Sally, and the other participants, solo motherhood threatened a child's 'need for a father' and there was much concern regarding how a donor-conceived child would feel about their family form. She continued:

> Throughout the whole thing my main concern, apart from the hundreds of other factors, has always been whether I am being selfish and whether a child, a teenager, an adult is at some point going to turn round to me and ask why did you do that? You've denied me knowing my father.

The participants expressed sadness that their child would not have a 'dad' and all described fathers as 'important'. Interestingly none of the participants mentioned the poorer child outcomes associated with single parenthood as one of their concerns about solo motherhood. However, they did feel that with only one parent their child may be 'missing out'. This was often described in terms of regret there would be no-one else who would love their child as much as them:

> And it's also a feeling that you belong to two people or at least that there is someone else who cares about you apart from your mother. Because everyone assumes that a mother's love is guaranteed. So yeah, I wish I didn't have to do this on my own. (Juliana, 39 years old)

All the participants described how men and women parented in different ways and how a balance of male and female influence would be beneficial to children. These differences often drew upon gendered stereotypes of parenting with the mother seen as a nurturer and the father as a disciplinarian engaging in 'rough and tumble' play.

The participants' concern regarding the absence of a father extended beyond the practical importance of a masculine influence, however, to a more symbolic importance that included *knowing* one's father (Graham, 2014). A child raised in a single parent family formed through separation or divorce would, in most cases, have contact with their biological father, if not be raised by him. If a child experienced the death of their father then they would have heard stories, seen pictures and videos of him, even if they were too young to remember him themselves. Even if conceived through a casual relationship the participants speculated that the mother would have some information, some memories of their child's biological father. However, a conception through donor sperm was constructed as a 'truly absent father': not only physically absent but also little, if any, information about him.

At the heart of the participants' concern about pursuing motherhood through sperm donation was fear that a lack of information about one's biological father would be detrimental to a child's sense of identity:

> I just thought I couldn't bring a child into the world where it doesn't have that; it will only have half the knowledge not the whole of it. I felt it would be difficult for the child not to be able to know half of itself. (Ruth, 47)

This fear seemed to stem from an understanding of identity that emphasises the importance of genetic ties: the idea that it is important for a person's psychological well-being to have knowledge of one's 'genetic origins' (Wilson, 1997). Such information was deemed essential for self-knowledge and understanding:

> Fiona: I didn't like that they wouldn't know where they came from
> Author: And why is that?
> Fiona: I don't know. I suppose that's difficult because I know where I come from so it's never been an issue for me. But, you know, I know I look like my dad, and I've got traits of both my parents and it just makes you feel secure I suppose. And I think when you are growing up and you've got friends who have mothers and fathers and if you haven't - well it just makes you understand who you are I suppose.

Indeed, it was contemplating the use of donor sperm that seemed to be a stumbling block for the participants: the greatest 'risk' in pursuing solo motherhood was its incorporation of donor conception.

Accounting for solo motherhood – being a 'good' mother

How did the participants move on from their conflicted and ambivalent feelings regarding solo motherhood? How did they reconcile becoming a single mother through sperm donation with their own constructs of being a 'good' mother?

Bock (2000) describes the US solo mothers she interviewed and observed in support groups claiming their legitimacy for motherhood through appropriating the term 'single mother by choice'. In contrast to my participants' ambivalence about the term, these American solo mothers felt the term depicted their legitimacy as mothers. Through being middle to upper class, well employed, politically aware and dedicated to motherhood, these single mothers argued for their place in the 'charmed circle' of the 'mother hierarchy', aligning themselves with the married, heterosexual and economically stable women who become mothers in their twenties and thirties, even though their marital status would generally preclude it. Bock (2000) describes how in doing so these women distinguish themselves from, and pass the stigma of single motherhood on to, teen and welfare dependent mothers. In contrast the participants in the current study seemed to align themselves with other single mothers when accounting for their route to motherhood. With single motherhood, albeit by circumstance, not 'choice', already accepted, if not a norm, within society, the participants were able to construct solo motherhood as a not too radical, and therefore not too problematic, departure from this existing social practice. Moreover, although their decision to puruse solo motherhood was perceived as a 'choice born out of constraint', the fact their journey to motherhood was 'planned and well considered' meant that the reality of being a single mother was expected and they would be better prepared for any difficulties they might encounter. Indeed, many of the participants commented that it was not just the single nature of their motherhood that was planned and well considered but their ideas regarding parenthood more generally. Making reference to friends who had 'just got pregnant' there was an assertion that they had themselves thought far more seriously about the implications of parenthood than the average heterosexual couple who just thinks, 'Lets get married and have a kid' without questioning 'a thousandth of what we have had to question'. The implication here seems to be that the more one has thought about, questioned and accounted for something, the more responsible one is for taking that action.

Furthermore, although the nuclear family as ideal, reiterated again and again in public and political discourse, had become an embedded norm for the participants, they were able to challenge the taken-for-granted rhetoric that a two-parent mother and father family is always best. Part of the normalising of solo motherhood as an option lay in the recognition that there were lots of different ways of forming a family. The rise in families headed by lesbian or gay parents helped the participants feel that they and their child would not be 'unusual'. Amanda described how going to an 'alternative family' event really helped her to see that solo motherhood could be an option for her, that one didn't just have to have the 'two up, two down, normal thing'.

In addition to seeing successful alternatives to the nuclear family, all the participants were only too aware that the nuclear family was not 'all rosy'. Many had direct experience of this: five of the participants' parents divorced; one participant's parents were both previously married with children and she grew up with stepsiblings; a further two participants experienced the death of a parent in childhood. Indeed, many participants used the rise in divorce rates to suggest that even if they were in the position of having a baby with a partner, there was a high chance they would end up a single parent anyway. Natasha elaborated:

> And when I think about all the families that have been torn apart through divorce or death or estrangement ... then I kind of think, well I could have been married for ten years by now, had two or three children, and been divorced by now so by default I'd still be a single mother.

Tellingly, Natasha was aware that this was her 'way of justifying' her decision to pursue solo motherhood. She continued:

> Ok, that is putting a very negative slant on marriage and that is quite likely my way of trying to convince myself that it would have happened anyway, trying to persuade myself that being a single mother by choice is going to be Ok.

Even if a relationship did not break down and a mother and father were parenting together, the participants were aware that they often looked at this 'idealised parenting set up' through 'rose tinted glasses'. Indeed, in recounting friends 'moaning' about their partners and husbands, the participants rationalised that partners were not always supportive and fathers not always involved in day-to-day parenting. Despite an era of supposed increased father involvement in parenting (Dermott, 2008) the participants wondered whether the time fathers spent with, and the influence they had on, their children was really that great:

> For some families, I'm not sure really, on a practical level day-to-day, if it makes a great deal of difference whether you are a single mother or a two parent family because fathers are so absent due to long hours at work. Most of the opposite sex couples that I know, where there are children, the father doesn't seem to be very involved in the children's lives. There's obviously affection between them but there doesn't seem to be very much involvement or interaction. (Claire, 40 years old)

Asserting the position that partners could actually be a 'hindrance' rather than help, thirteen of the participants claimed an advantage of pursuing solo motherhood would be that they wouldn't have a relationship to work at as well as a new baby to look after. Moreover, they would have full control over parenting decisions, doing what worked for them and the baby, rather than taking someone else's needs and opinions into account.

The participants did not view the absence of a partner as hindering the quality of parenting they could provide. References to being 'ready' and 'set up' for a child, having 'a big enough house and a car', as well as claims that 'I've had the expensive holidays, I've had years to do what I want' and instead being 'ready to settle down' and 'put the child first' were abundant in the interviews. Financial security was the attribute Bock's (2000) participants saw as key to justifying solo motherhood. Although my own participants stated that it was important to think about one's financial ability to raise a child alone, all asserted that this didn't mean you had to be 'rich' to pursue solo motherhood. Indeed there was huge variation in what 'being able to afford' solo motherhood meant: some participants had high salaries, equity in housing and large savings whilst others had little or no savings and would potentially go into debt with private fertility treatment. However, all believed they would be able to raise their child without resorting to social benefits. This was seen to be of utmost importance for external acceptance of their decision:

> So I think that as long as I can support myself and whatever family I have, as long as I am not seen to be grasping government hand outs or government accommodation then people couldn't care less what I do. (Natasha, 36 years old)

Access to social support was also an important consideration for the participants in contemplating solo motherhood. Support was described as a way to 'lessen the burden' and ensure a child wasn't 'missing out' by only having one parent. Indeed, it was the idea that their child would not be 'missing out' or 'lacking' in what they needed for a happy and content upbringing that seemed to be at the heart of the participants' accounts of why, despite diverging from parenting within the context of the nuclear family, they could still be a 'good parent'. In this regard a few of the participants described a sense of needing to go above and beyond providing a 'typical' upbringing. The 'lack of a father' was to some extent 'compensated' for through exposure to male role models, but they did not want a child, nor external others, to be able to claim they were 'lacking' anything else due to being raised in a single parent family. Karen explained:

> Well, there's the very simple things. Like, you know, I took a look at where am I in terms of my work, can I continue to work? Where am I in terms of my savings, my salary? Very practical things. Can I afford to give the child the best possible opportunities they need? Do I have a place to live? Do I have a support network? And then I did extra things on top of that, you know, what will private school cost? Again, it depends on what you need, but I guess if I'd been sitting here and I had a job that was not stable and I was in debt and I didn't have a place to live and I didn't have a support network; those things in my mind would've swayed my decision, I think. Because at the end of the day, when the child figures out that this is what you did and there's no real father there, what I don't want them to say is 'Oh, and because of that I didn't have this', or 'this happened to me', or 'that happened'. So I guess I stopped and thought about, OK, is there anything that will cause distress to the child, beyond the obvious, which is they may get upset that they didn't have, you know, Mummy and Daddy in a wedding car like their friend has?

It was the ability to be physically and emotionally available for their child, however, that was deemed to be most influential in one's ability to be a 'good mum'. Balancing being the sole financial provider for their family with being able to spend plenty of quality time with their child was therefore a concern for many of the participants. Although their form of employment varied, the majority of participants claimed they would be able to ensure some flexibility in working hours and be able to spend as much time as possible with their child. Indeed, even when only contemplating solo motherhood some of the participants started thinking about how they could improve the way their job would fit with children, considering, rejecting or embarking upon career changes, promotions and new opportunities. In doing so they seemed to be aligning themselves to fulfilling the growing ideal of 'intensive parenting' (Hays 1996): they may not be parenting within the nuclear family ideal and therefore accused of putting their own needs for motherhood above their child's need for a father but their desired child-centred approach to parenting meant that once mothers they would be engaged in sacrificing their own needs for the good of their child.

Donor conception – a gamble on imagined futures

The participants were able to reconcile becoming a single mother with being a 'good' parent. However, all participants struggled to incorporate the use of an unknown sperm donor into this conceptualisation: the 'risk' donor conception posed for their child was not so easily overcome. Indeed, when describing the decisions they made in pursuing solo motherhood and their thoughts pertaining to now undergoing fertility treatment, terms such as 'gambling', 'taking a leap of faith' and 'hoping' were abound in the interviews. At the heart of the use of such terminology was the belief that their child would one day judge their decision to become a mother through sperm donation. As Abby explained, 'I'm aware that it's something that I am going to be answerable to. It's a decision I'll have to be answerable for'. In trying to think about how their decisions would affect a child the participants were all too aware that they were 'taking a leap' into the unknown and 'gambling' on imagined futures. They did not know how their imagined child would react to the knowledge they were conceived via sperm donation:

> And I think also just the great unknown. It's a real great unknown of how they are going to deal with this. Will they be fine or won't they? It's just a jackpot really. And I don't know. (Jessica, 38 years old)

With the rhetoric of selfishness still permeating the participants' concerns regarding solo motherhood, Karen reflected upon the fact that she was prepared to pursue motherhood through sperm donation despite these unknowns remaining:

> Well of course I worry. And I think this is probably the only bit in this whole thing that I think potentially has a bit of an edge and a bit of selfishness. And yes I worry but I still want to do this badly enough that I will go ahead and do this anyway. I hope I can be the best mother I can and be supportive and make it a positive thing for them. But I can't guarantee that.

In their decisions regarding solo motherhood the participants seemed to be engaging in a process of 'risk management', or 'damage limitation': they did not see solo motherhood as an ideal scenario but with a mismatch of their ideals versus realities of family building they were trying to make solo motherhood as 'good as it can be' for their future child. Using sperm from identity-release donors in their treatment was one such 'damage limitation'. If having fertility treatment at a UK clinic the use of identity-release sperm donors was mandatory. However, if seeking treatment abroad the participants also had the option of using anonymously donated sperm: their child would never know more about their donor than the information included on his donor profile. The vast majority of participants felt identity-release sperm donation was preferable to anonymous donation: they believed that a child may be curious about their donor and should have the option of accessing identifying information and making contact with him at a later date (Graham 2014). The majority of participants felt that it was their child's 'right' to know this information:

> It has to be an open donor. That is absolutely key. My child has the right to know where they come from and I couldn't do it if that was not the case. The biggest factor for me is I'm going to be able to answer my child's questions. That whole thing about identity. That's been the biggest challenge. (Vicky, 37 years old)

Eleven of the study participants imported sperm from identity-release donors from European or American Sperm Banks for use in their treatment. The detailed donor information available about such donors was seen as one way of lessening the impact of the 'truly absent father' they felt a sperm donor represented (see Graham 2014). These participants were concerned that despite the removal of donor anonymity, the limited amount of donor information provided in the UK may leave their child curious about their donor, specifically about what he was *like*, before legally able to obtain further information and potentially meet him at age eighteen. The detailed donor information accessed through importing sperm meant these women could tell their child about their 'father' (Graham 2014):

> I'll be able to fill in quite a lot of blanks. I know he's a keen tennis player, I know the countries he's been to, I know his favourite food. Because I can paint that picture and create some character for a father figure it doesn't feel as bad as if I had only four lines of stats. (Abby, 41 years old)

Access to detailed donor information enabled these women to feel they *knew* the donor in some way. In turn they hoped their child would at least be able to know about, if not know in person, their biological father. They did not see solo motherhood as an ideal situation but hoped that through knowing detailed information about their donor, their child would feel comfortable with their donor-conceived origins and be able to form a positive self-identity: providing access to donor information was one way the participants could manage the 'risk' donor conception may pose for their child.

Conclusions

The participants were trying to make the best decisions they could given their current circumstances. Solo motherhood was a departure from their imagined life, a 'choice' made in the context of constraint, but one that they felt could still realise many of the aspects they deemed being a 'good' parent entailed.

Faircloth and Gürtin (2017) describe how paradoxically, while both ARTs and parenting expertise aim to 'assist' reproductive agents in their conception and child-rearing endeavours, they also generate new choices, burdens, responsibilities and accountabilities. The result is increasing anxiety and reproductive agents ending up feeling overwhelmed, scrutinised and 'not good enough' parents. This paper has shown how the prospect of solo motherhood provoked much anxiety and ambivalence for the participants in this study, with solo motherhood perceived as a 'risk' to the construct of a 'good mother'. As 'pre-conception parents' (Faircloth and Gürtin 2017) the single women considered, negotiated and accounted for the 'risks' solo motherhood may pose for their child: namely, being raised in a family departing from the nuclear family and not knowing their 'genetic origins'.

In contemporary society, engaging in risky behaviour has been described as a 'failure of self to take care of self' (Greco 1993, 357). Murphy (2000) describes the failure to live up to the ideal of the rational, responsible individual as inviting moral judgment. Such moral judgment, Murphy argues, acquires a particular force when the consequences of risky behaviour are borne not by self but others: special censure is reserved for those who knowingly put others at risk and the more involuntary the risk exposure, the greater the moral accountability of the risk taker. Hence Murphy describes how we read of 'victims' of drink driving or passive smoking. Mary Douglas suggested that to be put at risk is the equivalent of being sinned against, with the perpetrator of the risk being depicted as the sinner (Douglas 1990). As Murphy describes, the mother who 'chooses' to formula feed has to deal with the implicit (or occasionally explicit) charge that she has sinned against her baby (2000, 297). Likewise, the mother who 'chooses' to conceive a child into a single parent family, '*condemning* their child to a fatherless life' (Weathers 2007, my emphasis), is deemed at fault – in Douglas' term, to have sinned. As can be seen in the narratives of the twenty-three single women considering motherhood through sperm donation depicted in this paper, this 'sin' is discussed through the rhetoric of selfishness.

This paper has shown how the women in this study felt largely comfortable with their decision to pursue motherhood as a single person, believing themselves capable of fulfilling the requirements of 'good parenting' and negating many of the 'risks' growing up in a single parent household may pose. I argue that the cultural script of 'intensive mothering' as 'good mothering' may have been a factor in enabling these particular single women to reconcile solo motherhood with being a 'good' parent. Hays summarises the characteristics of intensive motherhood as 'child-centred, expert-guided, emotionally absorbing, labour intensive, and financially expensive' (Hays 1996, 8). Due to their class privilege, the single women in this study expected to be able to live up to this 'middle-class' ideal. The solo mothers whose narratives are portrayed in this paper are a self-selected group who were in the position of being able to afford private fertility treatment to achieve motherhood: even if they did not perceive themselves to be 'rich', they all identified as 'middle-class' and were in paid employment at the time of seeking treatment. Solo mothers from lower socio-economic backgrounds may have different concerns about becoming a solo mother and fulfilling the expectation of being a 'good parent'. However, regardless of their particular class and financial background, perhaps the physical absence of a male parent may remain relatively unproblematic for women considering solo motherhood due to the prioritisation of the mother-child relationship that 'intensive parenting' involves: 'The child-centred 'intensive' mother is one who is considered responsible for all aspects of her child's development – physical, social, emotional and cognitive – above and beyond anyone else, including the father (Hays 1996, 46). If 'good parenting' equates to 'good mothering' then the physical absence of a father in a solo mother family would remain a surmountable 'risk'.

However, there was no such cultural script for the solo mothers to draw upon when incorporating a sperm donor into their journey to motherhood. The solo mothers could not fully allay their anxieties regarding donor conception and the 'risk' this may pose for their child. Donor conception was the aspect of solo motherhood that the women remained feeling anxious about and accountable for. It is this particular feature of solo motherhood, I argue, where the participants became 'moral pioneers' (Rapp 1999), becoming 'philosophers of the private' whilst they did their best to employ strategies to align conceiving with donated gametes with being a 'good parent'.

The participants had tried their best to put their child's welfare at the forefront of their decision-making but were aware that the ramifications of their decisions were still unresolved. How would their child feel growing up in a household without a father? What would the child think about their decision to become a solo mother? How would they feel about being donor conceived? How would lack of knowledge about their biological father affect them as they grew up? However, the participants were pragmatic that they were not unique in entering into parenthood with concerns and anxieties: they believed having children and the experience of parenting was complex for everyone, and indeed was perhaps, to some degree, a moral experience of worry, concern and dilemma for any individual or couple. As Abby reflected, in deciding to become a parent everyone chooses their 'bucket of issues' on which to 'gamble':

> And then whatever will be will be. I guess it's kind of choose your bucket of issues. There are issues having a known donor, there are issues with having an unknown donor. There are issues with having an only child, there are issue with having twins. There are issues being a parent; if you don't want issues then don't be a parent. That's where I am kind of landed with all that.

Notes

1. A supportive network for families conceived with donated sperm, eggs or embryos, those considering or undergoing donor conception procedures and donor-conceived individuals. An email requesting participants for the research project was sent to all members of the single women's section of the DCN.
2. An online infertility community which has a forum for single women undergoing donor conception. A request for participants was posted on the 'Media and research requests' forum.
3. The London Women's Clinic is a private clinic providing IVF and fertility treatments to couples and individuals. It was one of the first clinics in the UK to treat lesbian and single women and is often described as the 'fertility clinic of choice for single women and lesbian couples'. Single women seeking treatment at the clinic were contacted by the sperm bank manager and asked if they would like to take part in the study.
4. Pseudonyms have been used throughout this chapter and any potentially identifying information has been removed or altered.

Ethical approval

The School of the Humanities and Social Sciences, University of Cambridge granted ethical approval for this study.

Acknowledgments

I would like to thank all the participants who generously gave their time to participate in the research that informs this article. I am grateful to the Wellcome Trust for providing the funding for the research and to Professor Susan Golombok and others at the Centre for Family Research, University of Cambridge, for their invaluable guidance and support during the research process. I would also like to thank the editors of this special edition and the two anonymous reviewers for their insightful comments and suggestions for the article.

Disclosure statement

No potential conflict of interest was reported by the author.

Funding

This work was supported by the Wellcome Trust [grant number 087238/Z/08/Z].

References

Almeling, R. 2011. *Sex Cells: The Medical Market for Eggs and Sperm*. Berkley: University of California Press

Amato, P.R. 2005. "The Impact of Family Formation Change on the Cognitive, Social, and Emotional Well-Being of the Next Generation." *The Future of Children* 15 (2): 75–96.

Beck, U. 1992. *Risk Society: Towards a New Modernity*. London: Sage

Bock, J. 2000. "Doing the Right Thing? Single Mothers by Choice and the Struggle for Legitimacy." *Gender and Society* 14 (1): 62–86.

Bristow, J. 2009. *Standing up to Supernanny*. Exeter: Imprint Academic.

Buttney, R. 1993. *Social Accountability in Communication*. London: Sage.

Carsten, J. 2000. *Cultures of Relatedness: New Approaches to the Study of Kinship*. Cambridge: Cambridge University Press.

Coleman, L. and F. Glenn. 2009. *When Couples Part: Understanding the Consequences for Adults and Children*. London: One Plus One.

Dermott, E. 2008. *Intimate Fatherhood: A Sociological Analysis*. London: Routeledge.

Douglas, M. 1990. "Risk as a Forensic Resource." *Daedalus* 119 (4): 1–16.

Dunn, J., K. Deater-Deckard, K. Pickering, T.G. O'Connor, and J. Golding. 1998. "Children's Adjustment and Prosocial Behaviour in Step-, Single-Parent, and Non-Stepfamily Settings: Findings From a Community Study." *Journal of Child Psychology and Psychiatry* 39 (8): 1083–1095.

Edwards, J. and M. Strathern. 2000. "Including our Own." In *Cultures of Relatedness: New Approaches to the Study of Kinship*, edited by J. Carsten. Cambridge: Cambridge University Press.

Faircloth, C. and Z. Gürtin. 2017. "Fertile Connections: Thinking Across Assisted Reproductive Technologies and Parenting Culture Studies." Sociology, 1–18. doi:10.1177/0038038517696219.

Franklin, S. and S. McKinnon. 2001. *Relative Values: Reconfiguring Kinship Studies*. Durham, NC: Duke University Press.

Giddens, A. 1999. "Risk and Responsibility." *Modern Law Review* 62 (1): 1–10

Graham, S. 2012. "Choosing Single Motherhood? Single Women Negotiating the Nuclear Family Ideal." In *Families Beyond the Nuclear Ideal*. edited by D. Cutas and S. Chan. London: Bloomsbury Academic;

Graham, S. 2014. "Stories of an Absent Father: Single Women Negotiating Relatedness Through Donor Profiles." In *Relatedness in Assisted Reproduction: Families, Origins and Identities*. edited by T. Freeman, S. Graham, F. Ebtehaj, and M. Richards. Cambridge: Cambridge University Press;

Graham, S., and A. Braverman. 2012. "ARTs and the Single Parent." In *Reproductive Donation: Practice, Policy and Bioethics*. edited by M. Richards, J. Appleby, and G. Pennings. Cambridge: Cambridge University Press;

Graham, S., S. Mohr, and K. Bourne. 2016. "Regulating the 'Good' Donor: The Expectations and Experiences of Sperm Donors in Denmark and Victoria, Australia." In *Regulating Reproduction*. edited by S. Golombok, R. Scott, J.B. Appleby, M. Richards, and S. Wilkinson. Cambridge: Cambridge University Press;

Greco, M. 1993. "Psychosomatic Subjects and the 'Duty to be Well': Personal Agency Within Medical Rationality." *Economy and Society* 22 (3): 357–372.

Hays, S. 1996. *The Cultural Contradictions of Mothering*. New Haven, CT: Yale University Press.

Hertz, R. 2006. *Single by Chance, Mothers by Choice. How Women Are Choosing Parenthood Without Marriage and Creating the New American Family*. New York, NY: Oxford University Press.

Inhorn, M. 2006. "'He won't be my Son': Middle Eastern Men's Discourses of Adoption and Gamete Donation." *Medical Anthropology Quarterly* 2 0 (1): 94–120.

Jadva, V., S. Badger, M. Morrissette, and S. Golombok. 2009. "Mom by Choice, Single by Life's Circumstance...' Findings from a Large Scale Survey of the Experiences of Single Mothers by Choice." *Human Fertility* 12 (4): 175–84.

Kirkman, M. 2004. "Saviours and Satyrs: Ambivalence in Narrative Meanings of Sperm Provision." *Culture, Health and Sexuality* 6 (4): 319–335.

Lee, E., J. Bristow, C. Faircloth, and J. Macvarish. 2014. *Parenting Culture Studies*. New York: Palgrave Macmillan;

Lee, E., J. Macvarish, and J. Bristow. 2010. "Risk, Health and Parenting Culture." *Health, Risk & Society* 12 (4): 293–300.

Mamo, L. 2005. "Biomedicalizing Kinship: Sperm Banks and the Creation of Affinity-Ties." *Science as Culture* 14 (3): 237–64.

Mannis, V.S. 1999. "Single Mothers by Choice." *Family Relations* 48 (2): 121–128.

Morgan, D. 2011. *Rethinking Family Practices*. London: Palgrave Macmillan.

Murphy, E. 1999. "Breast is Best': Infant Feeding Decisions and Maternal Deviance." *Sociology of Health and Illness* 21 (2): 187–208.

Murphy, E. 2000. "Risk, Responsibility, and Rhetoric in Infant Feeding." *Journal of Contemporary Ethnography* 29 (3): 291–325.

Murphy, E. 2003. “Expertise and Forms of Knowledge in the Government of Families.” *Sociological Review* 51 (4): 433–462.
Murray, C. and S. Golombok. 2005. “Going It Alone: Solo Mothers and Their Infants Conceived by Donor Insemination.” *American Journal of Orthopsychiatry* 2: 242–253.
Mohr, S. 2015. “Living Kinship Trouble: Danish Sperm Donors’ Narratives of Relatedness.” *Medical Anthropology* 34 (5): 470–484
Nordqvist, P. 2010. “Out of Sight, Out of Mind: Family Resemblances in Lesbian Donor Conception.” *Sociology* 44 (6): 1128–1144.
Pryor, J. and B. Rodgers. 2001. *Children in Changing Families: Life after Parental Separation.* Oxford: Blackwell Publishing.
Rapp, R. 1999. *Testing Women, Testing the Fetus: The Social Impact of Amniocentesis in America.* New York, NY: Routledge.
Richards, M., G. Pennings, and J.B. Appleby. 2012. *Reproductive Donation: Practice, Policy and Bioethics.* Cambridge: Cambridge University Press
Strathern, M. 1992. *Reproducing the Future. Essays on Anthropology, Kinship and the New Reproductive Technologies.* Manchester: Manchester University Press.
Wilson, S. 1997. “Identity, Genealogy, and the Social Family: The Case of Donor Insemination.” *International Journal of Law, Policy, and the Family* 11 (2): 270–97.
Weathers, H. 2007. “Who Needs Dads? The Single Mum’s Guide to Having a Baby Without a Man.” *The Daily Mail,* 29th October 2007. http://www.dailymail.co.uk/femail/article-497045/Who-needs-dads-The-single-mums-guide-having-baby-man.html

The role of normative ideologies of motherhood in intended mothers' experiences of egg donation in Canada

Kathleen Hammond

ABSTRACT
This paper explores 18 infertile Canadian intended mothers' experiences using donor eggs to conceive. Their narratives reflect a persistent normative ideology of reproduction and motherhood in Canada, where reproduction is natural, and expected, mother and child are genetically related, and two (genetic) parent families are what is normal. These ideologies permeate intended mothers' experiences of infertility, and their relationship with their egg donor. Much of the intended mothers' grief over their infertility was attributable to the gap between their experience and their perceived sense of normality. While intended mothers' relationships with their egg donor varied across a 'relational spectrum', their desire to minimize the gap between their experience and normative motherhood resulted in many intended mothers choosing less inclusive relationships, to the left of the spectrum, so as to mentally distance themselves from their use of donor eggs. Ultimately, however, in choosing to pursue egg donation, and in normalizing a different understanding of motherhood for themselves, these intended mothers are influencing cultural practices in Canada. In the long run, they, and others like them, are likely to alter cultural attitudes and conceptions of 'normality' when it comes to reproduction and motherhood.

Introduction

Egg donation was first used successfully in 1984 in Australia (Lutjen et al. 1984). It is one of a number of techniques, like sperm and embryo donation, surrogacy, and egg freezing that have provided new means for individuals to conceive in the decades following the development of in vitro fertilization and the birth of Louise Brown. Changing demographic and cultural trends have led to an increased reliance on assisted reproductive techniques like egg donation. More women are pursuing higher education, and participating in the labour force, and some are choosing to postpone childbearing (Statistics Canada 2017). Egg donation is also an important technique for family building among gay and single men (Richards, Pennings, and Appleby 2012). The Canadian Assisted Reproductive Technologies Register (CARTR) provides yearly data on the use of assisted reproductive technologies at the majority of Canadian fertility clinics. Review of yearly CARTR data, since its initiation in 2001, reveals an increase in the number of fertility clinics in Canada, and a general upward trend in the number of IVF cycles performed with donor eggs each year (Better Outcomes Registry and Network Ontario 2016; Gunby and Daya 2005).

Monica Konrad (2005, 9), who conducted research on anonymous egg donors and recipients in Britain, describes egg donation as a 'complex cultural phenomenon' because of the difficult nature of navigating the relations involved between an egg donor and intended mother. When donor eggs are used, an intended mother has no genetic link to the resulting child. Different from sperm donation, however, the intended mother gestates and gives birth to the child, and thus has a biological link to the child. As scholars such as Marilyn Strathern (1992) and Sarah Franklin (1997) have illustrated, the meaning of reproduction, kinship and relatedness, and the relationship between nature and culture are being re-crafted in the context of new reproductive technologies, and techniques. This, in turn, spills over into parenting culture, and in particular, cultural understandings of what makes a parent. The connection between egg donation and parenthood in the Canadian context is explored here as part of an effort, initiated by Charlotte Faircloth and Zeynep Gürtin (2017), to develop a sustained dialogue between the areas of 'assisted reproductive technologies' and 'parenting culture'.

How do ideologies of reproduction, kinship, and the relationship between nature and culture manifest in normative ideologies of motherhood in Canadian society? What role do normative ideologies of motherhood play in shaping the experiences of infertile Canadian women using donor eggs to conceive? This paper explores these questions through the narratives of 18 Canadian intended mothers. These women are confronted with normative ideologies of reproduction and motherhood where reproduction is natural and expected, mother and child are genetically related, and two (genetic) parent families are what is normal. I illustrate the ways in which these ideologies permeate intended mothers' experience of infertility and their relationship with their egg donor. Intended mothers' concerns over deviating from what they consider to be 'normal' reproduction and motherhood contribute to their grief over a diagnosis of infertility, and influence their preference for less inclusive relationships with their egg donor. Ultimately, however, in choosing to pursue egg donation, and in attempting to normalize a different understanding of motherhood for themselves, these intended mothers play a role in influencing cultural practices. In the long run, they, and others like them, are likely to alter cultural attitudes, and perhaps widen the persisting narrow conception of normality when it comes to reproduction and motherhood.

Infertility and normality

Beginning with the earliest studies on infertility, sociologists and anthropologists have portrayed how women experiencing infertility, and using assisted reproductive technologies, have grappled with the ways that their experiences differ from their perceived sense of normal reproduction and motherhood. This begins with our cultural expectations about the courses we anticipate our lives to take, including expectations of whether and when to marry and to have children. In Western societies, these cultural ideas of life course are that it be 'orderly, predictable, and continuous' (Becker, 1997, xxi). As Gay Becker illustrates in 'The Elusive Embryo', a diagnosis of infertility disrupts this life course, and confronts assumptions of normality, as it is considered to be abnormal to be unable to reproduce biologically. Franklin (1997) argues that for the infertile women in her study, the discovery of infertility opened up a pervasive and symbolic cultural gap, separating them 'from the life they expected to lead' (139). Sandelowski (1993) conveys the ways in which the real and perceived stigma of infertility pervades the lives of infertile people, women in particular, and spoils their identities. Sandelowski (1993) found that infertile women drew an amorphous boundary between normal and abnormal—between those who are able and unable to have children.

Thompson's (2005) ethnography of the daily interactions of IVF clinics demonstrates how through dynamic coordination—the 'ontological choreography' of fertility clinics—fertility clinic staff help to try to close this gap between 'normal' and 'abnormal' by making the technologies feel normalized, even natural (8). Through use of IVF, Franklin (1997) illustrates how her participants' understandings of conception changed. The view of conception within assisted reproduction is that nature is a badly designed machine and that technology is simply giving 'nature a helping hand' to help it do what it would do anyway (188). Having a child without technological assistance, like IVF, would be 'miraculous' (188). Thus, these women, and the fertility clinic staff, try to find a way to make the use of these technologies feel normal.

Egg donation and ideologies of motherhood in Canada

Needing donor eggs raises particular challenges because it means that intended mothers will not have a genetic link with their resulting child. Genetic links have been found to be significant to kinship models (Becker, Butler, and Nachtigall 2005; Nelkin and Lindee 1995; Ragone 1994; Schneider 1984). In Canada, the importance of genetics to ideologies of motherhood was evident through intended mother's narratives. It is also evident in Canadian law. For instance, the parental status of the non-genetically related mother in same-sex relationships in Canada has often been precarious. It has only been in recent years that a number of provinces, including British Columbia (2014) and Ontario (2016), have passed legislation making it so that same-sex parents do not have to adopt their own children. Under Canada's *Citizenship Law Act*, 1985 children born outside of Canada acquire citizenship through descent or parentage if at least one of their parents is Canadian. Canadians who do not use their own sperm or eggs to conceive cannot pass on their citizenship to their children (Carsley 2016). The *Assisted Human Reproduction Act*, 2004, the legislation that regulates egg donation in Canada, allows egg donors to be reimbursed for expenses but not compensated. This has made it very difficult to find willing egg donors, and particularly difficult to find donors who are not a relative or friend. Thus, the Act seems to inadvertently encourage donation from someone with whom an intended parent already has a biological link, like a sister. A genetic link is then maintained between intended mother and child.

In 'Nameless Relations', Monica Konrad (2005) illustrates how the egg recipients she interviewed tried to overcome the lack of genetic link. Konrad observed that her participants emphasized their relatedness with offspring conceived through donor eggs because of the ties they forge through the gestational and birthing experience, which they privileged on the 'biological continuum of relatedness' (156). Konrad (2005) additionally found that intended parents were concerned with being matched with donors who looked like them, or their husband, so that their child would physically resemble them. Genetic links between children and their parents are often commented upon by others through what Becker, Butler, and Nachtigall (2005) term 'resemblance talk' (1301). Intended parents want their child to look like them, so that the child appears to be their 'natural own'. This way, they can avoid possibly stigmatizing comments.

In terms of recipients' feeling about the egg donor, Konrad (2005) found that they varied. Egg recipient Wendy, for example, described her egg donor as someone she would never forget, while other egg recipients tried to 'cancel out' the donor (143). At the time that Konrad (2005) conducted her study, egg donation in Britain was altruistic. Konrad (2005) found that donors and recipients alike framed the donation as a 'gift'. The debate over framing has been well-rehearsed in the literature. In particular, much attention has been given to the framing of donation as a gift versus as an economic transaction. Scholars, like Spar (2006) for instance have suggested that it is the exchange of money, or not, that structures the transaction. Almeling (2009), however, explored the different embodied experiences of American egg and sperm donors. Although both were paid, she found that egg donors saw donation as a gift, while sperm donors perceived it as a job. According to Almeling (2009), this is because the way that the transaction is framed is just as fundamental to structuring the exchange as the money changing hands. The difference between sperm and egg donors' understandings of donation has to do with the different and gendered ways that donation is framed by egg agencies and sperm banks. In the Canadian context, where payment for eggs is not permitted, and a gift rhetoric often surrounds donation, Cattapan (2016) argues that we should reconceive of egg donation as labour. In line with Almeling's (2009) reasoning, Cattapan seems to suggest that allowing payment will not be sufficient to restructure the practice of egg donation. Rather, to better address the interests of Canadian egg donors, we need to re-frame the practice of egg donation altogether.

Methods

Between October 2013 and March 2015, I collected the data for this study through qualitative interviews with 18 Canadian intended mothers. Fourteen intended mothers self-identified as Caucasian, two as African Canadians, and two as Chinese Canadians. Intended mothers were a well-educated and largely affluent group, and had a wide range of occupations, including medical doctor, lawyer, university professor, chef, dental hygienist, homemaker and small business owner. Two intended mothers self-identified as lesbians, the rest described themselves as heterosexual.

Of the intended mothers with whom I spoke, four were pregnant, three had children conceived through the use of donor eggs, and 11 were still in the process of donation. In Canada, there are no restrictions on donating anonymously. Eleven intended mothers had anonymous donors, and so had little or no contact with the donor. Four intended mothers had an identified donor, meaning that they met the donor through the process of donation and stayed in contact with her after the donation. Three intended mothers had a known donor—either a family member or friend. All intended mothers had pursued their preferred type of donation, aside from two who had had anonymous donors, because of being unable to find identified donors.

A small body of research has documented the trend of Canadians seeking eggs in other jurisdictions, usually where donors are paid (e.g. Blyth 2010; Hudson et al. 2011). This study reflected those findings, and intended mothers in this study had pursued paid donation in a variety of countries, including the USA, Mexico, Barbados, and Greece. All intended mothers who had pursued donation in countries outside of Canada had compensated their donors, or planned to. Although the *Assisted Human Reproduction Act*, 2004 bans payment, it is permissible to reimburse donors for expenses incurred through the donation. However, since what constitutes reimbursable expenses has never been clarified, it is a grey area in which the various players have been left to interpret the law themselves. All intended mothers who worked with donors in Canada had reimbursed their donor's expenses. Many had reimbursed their donors in such a way that, depending on your interpretation of the Act, could constitute compensation. In their view, however, it was reimbursement. Certainly, all intended mothers in this study unanimously thought that egg donors should be paid. Detailed information about the participants, their age, and status at the time of interview, as well as information about where they pursued donation can be found in Table 1 at the end of this section. All participants in this study have been given pseudonyms.

In Canada, infertility forums and advertising websites have become a popular way for intended parents and egg donors to connect. I recruited intended mothers through seven of these advertisement and infertility forums. These were kijiji, Craig's List, Toronto Super Ads, opts.com, IVF.ca, ivf-infertilty.com and co-parent-search.com. I posted information about this study on all seven forums, and invited intended mothers who had posted in these forums to participate. A Canadian infertility support group also kindly circulated information about this study.

I conducted phone and Skype interviews. Phone and Skype were preferred because intended mothers lived in many different parts of Canada, and because they afforded intended mothers a greater sense of anonymity. This was important given the private nature of infertility. I asked open-ended questions about their decision to use donor eggs, logistical details about how they found an egg donor, their feelings toward their donor, and, finally, their perspectives about the laws in Canada surrounding egg donation. Intended mothers were interviewed once, and interviews ranged in length from half an hour to just over two hours, with an average of 74 minutes. I transcribed and coded all interviews according to conventions outlined by Foss and Waters (2011) and Ryan and Bernard (2000).

I present the data in two sections. The first, 'Coming to Terms', reveals intended mothers' reactions to being told they would be unable to conceive with their own eggs and their normalization of a different conception of motherhood. In the second section, 'Pursuing Egg Donation', I explore the range of relationships between intended mothers and egg donors along what I term a 'relational spectrum'.

Coming to terms

All of the intended mothers but one—who had proposed egg donation to her physician—had had egg donation suggested to them by a physician. Prior to reaching this juncture, all had experienced futile attempts with various other fertility treatments. Intended mothers described their grief and shock over the news, many having remained optimistic that they would eventually conceive with their own eggs. Natasha shared her experience of being given the news. She said, 'It was just kind of like having your loved one get hit by a car on the street, and they are dead. And it's unexpected. And literally, like, I really felt that a part of me has died.' The ability to become biological parents is assumed – it is, after all, the norm. For many women and men, this ability is strongly linked with their identity. Threat to the ability to reproduce comes as a threat, or demolishment, of part of their identity,

Table 1. Intended mother profiles.

Name	Age when seeking an egg donor	Marital status	Status at time of interview	Type of egg donation	Country where pursued/ pursuing egg donation
Jessica	40	Married	Pregnant and a few weeks from due date.	Anonymous	USA
Monique	38	Married	Trying to find a donor in Canada. Has frozen eggs from a USA agency, and completed an unsuccessful fresh donor cycle with a Canadian donor.	Identified	Canada and USA
Natasha	44	Married	Pregnant and a few weeks from due date. Had also used donor sperm. One previous unsuccessful donor cycle.	Anonymous	USA
Jennifer	34	Married	Pregnant and a few weeks from due date.	Anonymous	USA
Maria	43	Single	In the midst of a donor cycle. Also using donor sperm.	Identified	Canada
Amy	31	Married	Trying to find a donor.	Identified or anonymous	Mexico
Leah	42	Married	In the midst of a second donor cycle. One previous unsuccessful donor cycle.	Identified and known (sister-in- law)	Canada
Lucie	33	Married	In the midst of a first donor cycle.	Known (sister)	Canada
Valia	40	Married	Seeking a donor in Europe.	Anonymous	Greece or Spain
Anna	42	Single	Seeking a donor in Barbados. Trying to conceive with donor eggs and sperm. Three prior unsuccessful cycles in Canada.	Anonymous	Canada and Barbados
Iva	43	Married	In the midst of a first donor cycle. Two children from a previous marriage.	Anonymous	Greece
Rebekah	40	Married	Starting a donor cycle. One previous unsuccessful cycle.	Anonymous	USA
Evelyn	41	Single	Has a nine-year old daughter conceived through donor eggs and sperm.	Anonymous	Canada
Cheryl	42	Married	Has six year-old twins, and a four year-old, all conceived through donor eggs.	Anonymous	Canada
Kyra	29	Married	Has a four-year-old daughter. Now looking for a donor to have a second child.	Known (friend)	Canada and USA
Lyndsey	41	Married	13 weeks pregnant through egg donation.	Anonymous	USA
Emily	42	Married	Just connected with a donor, and about to start a first donor cycle.	Identified	Canada
Kara	43	Single	Has a two-year-old daughter conceived through donor eggs and sperm, and about to undergo an embryo transfer with remaining embryos from having her daughter.	Anonymous	Canada

hence 'part of me has died.' This is particularly the case for women (Firestone 1970; Pfeffer 1987; Rapp 2000). The feeling of grief was felt unanimously among intended mothers, although there were some differences in what made the news particularly grievous.

Six intended mothers spoke in depth of their feelings that they were not making their biological contribution to the child. Monique said:

> It was the thought of having someone else's eggs in me. It was that weird feeling that I had to, you know, use someone else's eggs because I was deficient. It was sort of you know, um, that weird concept of having this foreign object in me, but the factor that I was probably slamming myself at the same time because I couldn't conceive. It was me who was infertile and I had to get help. So I didn't, I didn't like that.

Eggs are 'cultural icons of gender' and epitomize cultural ideals of womanhood (Becker 2000, 133). Monique describes herself as being 'deficient'. Her eggs being unviable undermines her cultural understandings of her womanhood, and as a result, her role as a mother. Intended mothers expressed frustration and anger over their eggs being unviable and described feeling 'cheated' and 'let down' because of their inability to use their own eggs, and to reproduce in the 'normal' way. Alongside this frustration over the inability to use their own eggs, was, for some intended mothers, a discomfort with using someone else's eggs. Monique, for instance, notes the unusualness of using someone else's eggs and her unease with having to do so.

The most significant focal point for the grief and sadness described by intended mothers was the lack of genetic connection that they would have with their child. Anna expressed sadness for her mother, that her mother would not have a genetically related grandchild. Intended mothers, Maria and Natasha, discussed their desire to look down into the face of their future child and see themselves. Our conversation suggested that their grief was over the perceived disruption to generational continuity of their family, because of the lack of genetic connection. 'I was disappointed that I would not have a biological child and worried that the child would not look like me which might cause people to constantly ask me questions about it', said Anna. Anna indicates her concern about other people's perceptions and reactions to what she perceives to be abnormality caused by a lack of genetic link with her child. Intended mothers' sadness over a lack of generational continuity, lack of physical resemblance, and how this lack of genetic link might be perceived by others, however, indicates that their sadness is not necessarily based in the literal loss of the genetic link, but rather in their loss of what this genetic link symbolizes—typical motherhood.

Jessica took some consolation in the hope that by carrying the child she might come to see herself as a contributor. She said:

> Maybe not DNA, but I am a contributor. For me, the most important thing was to never look at my husband and feel as though he had contributed something I had not. And to never feel offensive, uh, lesser, I don't even know how to put it into words anymore. To feel a sense of, a lack of contribution.

Not only does Jessica feel remorse over not having a genetic link with the child, there is a sense of not having contributed something that her husband has, as though he has fulfilled his parental role, but she has not fulfilled hers. Kyra also spoke to this apprehension. She said, 'I think I felt insecure about like, worried that my child and my husband would have like this really strong connection and I would be the outsider in the family.' Here, Kyra illustrates her fear of a lack of resulting connection with the child because of the lack of genetic link. Genetic connections, which are central to normative ideologies of motherhood, are perceived to be the basis of the specific nature of the parental-child relationship. Additionally, Kyra's fear is exacerbated by her concern that her husband will have the connection that she would not, since he will be genetically related to the child.

Intended mothers' reactions to the news that they would have to use donor eggs are replete with references to a normative ideology of motherhood. The reasons for their grief reveal assumptions about the nature of this normative ideology. Their grief and their worries about conjugal inequality, and how their experience will be perceived by others, seem to result from the gap between their experience and what they perceive to be normal. Coming to terms with this reality, and deciding to go forward with using donor eggs was difficult.

The decision to proceed was gradual, ranging from a number of months to a number of years. Monique, for instance, sat on the idea of donor eggs for four years. During this transition time, intended mothers grieved, and 12 looked into adoption. Part of deciding to take the egg donor option involved intended mothers re-thinking and normalizing a new conception of motherhood—one in which their child would not be genetically related to them (Becker 2000; Thompson 2002). Intended mothers, like Anna and Maria, normalized motherhood, where a mother does not have a genetic link with their child, by referencing mothers they knew who were not genetically related to their children. 'Biological [is] not the end all and be all about being a mother', said Anna, reflecting on the relationship that her sister-in-law has with her adopted daughters. Maria said:

> I have many friends who've adopted children, and I don't think of their children as not their children because she, you know, she was born in Korea and my friend wasn't. Do you know what I mean? So rationally when it all came together, I was like, I just want the end. And I'll do whatever steps I need to get there.

Both Anna and Maria indicate their desire for the 'end'—the experience of having and raising a child. They were both willing to forego the genetic link in order to have that experience. Ultimately, all of the intended mothers, with whom I spoke, decided to go forward with egg donation, having decided that they did not need a genetic link with their child to be a mother.

Pursuing egg donation

Having decided to pursue egg donation, intended mothers' decide where to pursue donation (either in Canada or abroad) and what type of donation—known, identified, or anonymous, to pursue. The relationships between intended mothers and egg donors fell into three segments across a relational spectrum. These three segments were defined by 'distance and cancelling out', 'acknowledgment and gratitude', and 'contact and intimacy'. As will become apparent, the type of donation that intended mothers chose to pursue in some ways dictated the nature of the relationship they would have with the egg donor. An anonymous donation, for instance, is unlikely to be characterized by 'contact and intimacy' because the anonymous nature of these donations means that there is no contact between intended mothers and their donor. Donation from a friend or family member, however, is unlikely to result in a relationship defined by 'distance and canceling out' because of the difficulty of canceling out a friend or family member who is present in an intended mother's life. However, donation from a friend, or family member, could be characterized by 'acknowledgment and gratitude.' Intended mothers in this study were at varying stages of the donation process. Thus, at times intended mothers' observations were anticipated, or reflective, depending on what point they were at in the donation process at the time of interview. I indicate where intended mothers were in the process.

'Distance and cancelling out'

To the left side of the spectrum are those relationships characterized by distance and cancelling out. All of these intended mothers had had anonymous donors. The majority of these intended mothers had pursued donation abroad, and all of those who had, had paid their donor. Those intended mothers who had pursued donation in Canada had generously reimbursed their donors. These intended mothers tried to distance themselves, as much as possible, from the donor, almost as though to erase their use of donor eggs. Jessica was pregnant at the time we spoke. In describing her relationship with the anonymous egg donor, she said:

> Um, a lot of people say, you know, my donor is my angel. I don't know that I'd go that far. I might be a little more callous, well not callous, but I don't consider her my angel. I consider what happened a business transaction. I know that probably sounds really weird. Again I'm thankful for the aspect that she put her body through potentially dangerous things. Um, that is awesome. But she was reimbursed and she was reimbursed well. I look at it more as a business transaction. I know that probably sounds a little cold but maybe that's my coping mechanism as well, to not put a personal twist on it.

The egg donor represents Jessica's discomfort with not having been able to use her own eggs to conceive. Perhaps an additional source of discomfort, Jessica also acknowledges that the egg donor is taking a risk in order to help her fulfill her goal of becoming a parent. As a coping mechanism, Jessica frames the donation as a 'business transaction', allowing her to limit the relationship to the moment of the exchange, and maintain distance from the egg donor. This helps to lessen the gap between Jessica's route to motherhood and normative ideologies of motherhood. Jessica also stresses that the donor was 'reimbursed' and 'reimbursed well.' Payment serves as a way for intended mothers to feel more ethical about the donation, because the donor receives something in return for the donation. Having received the eggs and paid the egg donor, the relationship is over. In line with maintaining distance from the egg donor, Jessica did not inquire or involve herself in the donor's experience of the donation. She did not know, for instance, that her donor had experienced hyper-stimulation. 'The clinic just mentioned it when we were clearing up the rest of our account, that she was hyper-stimulated and that she required some medication and we would have to pay for it. So we did.'

Monica Konrad (2005) observed that some of the egg recipients she interviewed 'cancelled out' the egg donor, and their use of her eggs. Like those egg recipients described by Konrad (2005), Jessica also seemed to have been able to cancel out the egg donor to some extent. A month away from her due date when we spoke, she shared with me that she often forgot about the donor altogether. She said:

> I mean a couple of weeks ago my husband said to me 'oh my god I forgot that we even had a donor.' I mean we were just talking about something random to some random person, and

> my husband was like 'oh my god I totally forgot.' And that was really nice. And I know that I forget all the time too.

Another intended mother, Maria, was in the midst of trying to find a donor. 'My goal', she said, 'is to feel as little for them [the egg donor] as I can.' Like Jessica, disconnecting from the egg donor was a way for Maria to disconnect from her inability to use her own eggs. Maria, was, however, hoping to leave the option open for any child she had to be able to contact the donor in the future if the child wished. She made it clear, however, that this did not mean that she wanted a relationship with the donor. She said:

> I don't want to have an ongoing relationship with the person. I think that overcomplicates matters. For me anyway, emotionally. This isn't an open adoption. All I want is that if I ever have a reason medically, or if my child when they're of legal age at 18 want to know where their genetic heritage is from, that they could find the person. That's all I'm interested in. I'm not interested in having annual meetings with the person.

Of all of the intended mothers with whom I spoke, Evelyn was perhaps the most far removed from the donation. She had a nine-year-old daughter, and had gone through the donation almost 10 years prior to our interview. She said: 'And I find that it's kind of painful for me to think a lot about the egg donor because that's just for me psychologically it's sort of about my inability to conceive myself and use my own eggs.' Reflecting on her attitude towards the donor, years prior, Evelyn said:

> I'm sure I was less concerned about the risks for the egg donor compared to the risks that I took when I was going through all the fertility treatments. I don't think I had a lot of empathy at the time for the risks that the egg donor would have been taking because I was so focused on becoming a mom.

Evelyn is reflecting on her lack of reflection 10 years prior. She was so consumed with becoming a mom that she gave little regard to the experience of the egg donor. Part of this drive might have resulted from the feeling of needing to fulfill this expected aspect of her identity. Additionally, as she points out, thinking about the egg donor is painful, and so perhaps, like Jessica and Maria, part of not thinking about the risks had to do with trying to cope.

In line with trying to distance themselves from or cancel out the egg donor, these intended mothers only shared their use of donor eggs with a small number of people, if at all. All the same, they play a role in influencing cultural conceptions of motherhood. They are doing so, firstly, through exposing others to egg donation by having participated in this research, and by sharing their story with others – no matter how few people they share it with. Some of these people with whom they share their story might, in turn, share it with others, which will lead to more people being exposed to the use of egg donation. Thirdly, by using egg donation many of these women become part of the growing statistic describing the number of people using donor eggs nationally that is published, for instance, by CARTR. For people considering using egg donation, looking at these numbers and seeing the growing number of people becoming parents through egg donation helps to normalize this type of motherhood, and might also encourage it.

'Acknowledgment and gratitude'

Somewhere in the middle of the spectrum, the relationship shifts from being characterized by 'distance and cancelling out' to being characterized by 'acknowledgment and gratitude'. All of the intended mothers whose relationships fell into this part of the spectrum had anonymous donors. Many had pursued donation abroad, and all of those who had, had paid their donor. Those intended mothers who had pursued donation in Canada had generously reimbursed their donors. This second part of the spectrum was distinguished by an enduring link, of indebtedness or gratitude, that intended mothers felt they had with the donor. These intended mothers were more willing to acknowledge the role of the egg donor. All of these intended mothers had shared their use of donor eggs with a handful of family and close friends. However, they still had no, or little contact, with the donor during the donation, or following it.

Natasha, an intended mother who was several weeks pregnant, described feeling 'grateful to have just had the opportunity.' 'I think I will always feel connected to her', she explained. Another intended mother, Valia, said, 'I don't just see the donor as a vending machine. How can I ever repay her for her incredible gift? I anticipate feeling linked for life to that donor'. Valia had not yet gone through a donation, but planned to pay the donor. Like the egg donors interviewed by Rene Almeling (2009), the exchange of money did not seem to be the only factor in structuring the transaction. These intended mothers framed the donation more like a 'gift'. Although they still seemed to want to maintain distance with the donor, they were more willing to acknowledge her.

Two of the intended mothers, Rebekah and Iva, had left gifts for their donors in addition to having paid them. All of the intended mothers made efforts to inform themselves about the procedure and risks for the donor. In deciding which clinic or agency to work with, these intended mothers had put some consideration into the clinic or agency's reputation for treating their donors well. Iva and her husband, for instance, had looked carefully into clinic protocols for donors before deciding where they would undergo the donation. Iva had had an anonymous donation, arranged by her clinic. The clinic was limited in what they could share with Iva. Regardless, Iva followed-up with the clinic after the donor's cycle to see how it had gone.

These intended mothers' acknowledgment of the donors was also evident in the way that they spoke about their connection with the donor. Natasha and Valia, for instance, in noting an 'enduring connection' with the donor, are acknowledging the contribution of the donor in creating their child. Jennifer, who was pregnant at the time of our interview, said:

> As someone who's now almost half way through my pregnancy with a donor egg, it so very rarely crosses my mind that the child that I'm carrying is genetically not related to me. Like it's really just, once that embryo was in me, and implanted in me, and confirmed as a pregnancy, that's now my kid. And the donor did her deed up front, and we're thankful because this child that we're creating, this special little human being, is due in part to her. From that point where she did her donation, and from the point where that embryo was implanted in me, now I'm the one doing the job.

When Jennifer was first given the news that she would need to use donor eggs, she worried about the lack of genetic link she would have with a child conceived through donor eggs. Like Jessica, who is described in the 'distance and canceling out' section as often forgetting about the donor, Jennifer also notes that she rarely considers that the child is not genetically related to her. For Jennifer, however, this is not because of forgetting about the donor. Like the egg recipients described by Konrad (2005), through gestating the child, Jennifer 'took over' from the donor. In doing so, the child became 'her kid'. This is what made it easy to forget that the child is not actually genetically related to her. Intended mothers like Jennifer, who share their experience, do not only expose others to egg donation but expose them to a comfort with the practice.

'Contact and intimacy'

To the right of the spectrum are those relationships defined by 'contact and intimacy.' All of these intended mothers had worked with donors in Canada, and their donors had been identified or known. They had all reimbursed their donor's expenses, and for the most part they had been conservative in their reimbursements. These intended mothers were very involved in the donor side of the process. They had not only looked into the process, they also took steps to ensure that their donor was informed, and certain about their decision to donate. Emily who was about to begin donation with an identified donor, had sent information from her local infertility support network 'about basic requirements and other information, about risks, etc.' to all of the egg donors with whom she had spoken, including the woman who ultimately donated to her. Leah, whose sister-in-law Lisa offered to donate to her and her husband, said: 'We gave Lisa ample opportunity to change her mind after she had gained a full understanding of the process. We wanted to be sure she knew what she was offering, in terms of time, etc.' Monique had offered to accompany her identified donor to the retrieval process. She said:

> I think I was probably more stressed during the donor cycle then I was during my own cycle four years ago. Because it's someone else doing that for you, rather than if you were doing it to yourself. She's taking on the risks, not you. So, it's, I found that probably the hardest.

Monique also followed up with the donor several times after the retrieval. She acknowledges that the egg donor is taking risks in helping her to fulfill her goal of motherhood.

Three of the intended mothers, whose relationships fell in this part of the spectrum, had undergone known donations. Leah's donor had been her sister-in-law. Even though the donation had been unsuccessful, Leah remarked that the experience had brought them 'closer than ever'. Another intended mother, Kyra, had successfully had a child through the donation. Kyra's donor was a good family friend. In describing her relationship with the donor, she said:

> I send pictures and we're really good friends and we talk every week kind of thing. Yeah, I mean she'll always be a part of our life right? She'll always be a part of our life and luckily we all get along really well and we're all like family.

Kyra shared that sometimes it was difficult navigating her close friend being the genetic parent to her child, but largely, as her quote suggests, it is a positive experience. Kyra, Monique and Leah had all shared their use of donor eggs with many family members and friends, many of whom as a result witnessed these women navigate and play role in normalizing a new ideology of motherhood, and family.

Conclusion

In the decades since the development of IVF, the technology and medical procedures used to treat infertility have continued to develop and change. As Gay Becker (2000) contends in 'The Elusive Embryo', however, the source of people's grief over a diagnosis of infertility has not. At the heart of this grief are people's concerns over the ways in which their experience differs from what they perceive to be 'normal'. The intended mothers in this study, for instance, were very concerned with the fact that they would not have a genetic connection with their child. Their comments suggest, however, that their grief stemmed not so much from the literal loss of the genetic link, but from losing the opportunity to experience typical motherhood.

Intended mothers' concerns over the gap between their lived experience and the normative ideology of motherhood in Canada was evident in the spectrum of relationships that emerged between them and their donors. For instance, most intended mothers had worked with anonymous egg donors (both in Canada and abroad) and had either paid or generously reimbursed their donors. It might be expected that these relationships would look relatively similar. However, this was not the case. While some of these relationships were more transactional and could be characterized by 'distance and canceling out', others were better defined by 'acknowledgment and gratitude.' Intended mothers whose relationships with their donor fell into the latter category had looked into clinic procedures, left gifts for their donors, and acknowledged an enduring link between them and their donor. They shared these characteristics with the small group of intended mothers who had worked with identified or known donors in Canada, had conservatively reimbursed their donors, and whose relationships were characterized by 'contact and intimacy.' The differences between the 'distance and canceling out' and 'acknowledgment and gratitude' groups, despite donors in both groups having been paid, supports Almeling's (2009) findings that money alone does not structure the nature of these exchanges.

In this study, intended mothers' feelings about deviating from cultural understandings of motherhood seemed to play an important role in the way that they structured their relationships with egg donors. All intended mothers expressed discomfort with deviating from these norms, and had concerns about navigating their relationship with their egg donor. However, intended mothers who seemed more uncomfortable with it had relationships with their donor that were characterized by qualities farther to the left of the spectrum. They were more prone to distance themselves from or 'cancel out' the egg donor. Think back to Evelyn, for instance, who explained that she did not put a lot of thought into the risks for the egg donor because thinking about the egg donor is painful as it is a reminder of her inability to use her own eggs. As found in previous studies of egg donation [e.g. Almeling (2009) and Monica Konrad (2005)], intended mothers employed a particular rhetoric to describe the nature of the transaction, for instance some described it as a 'gift' or as a 'business transaction'. Here, intended mothers seemed to use these terms as a means to explain or justify the type of relationship they had with the donor.

Certainly, there is no right or wrong or better or worse way for intended mothers to structure their relationship with their egg donor. What these findings highlight, however, is the ways in which intended mothers' concerns over deviating from narrow normative ideologies of motherhood make the experience of infertility more painful, and their decisions around whether and how to pursue egg donation more complicated. Little by little, intended mothers, like the ones who participated in this study, are likely to alter cultural attitudes by using assisted reproductive techniques and exposing others to egg donation. If this occurs, then perhaps eventually the source of the anguish of infertility and force behind much decision-making in the course of using techniques like egg donation will no longer be people's concerns over deviating from 'normality'.

Ethical approval

This research was approved by the Department of Sociology of the University of Cambridge, and was conducted in compliance with departmental guidelines.

Acknowledgments

The author firstly wishes to thank the intended mothers who so generously shared their stories and insights. This article benefited enormously from feedback from Zeynep Gürtin, Charlotte Faircloth, and two anonymous peer reviewers. This research was made possible by the Gates Cambridge scholarship and the Cambridge Commonwealth Trust.

Disclosure statement

No potential conflict of interest was reported by the author.

References

Almeling, R. 2009. "Gender and the value of bodily goods: Commodification in egg and sperm donation." *Law and Contemporary Problems* 72:37.

Becker, G. 1997. *Healing the Infertile Family*. Berkeley, CA: University of California Press.

Becker, G. 2000. *The Elusive Embryo*. Berkeley, CA: University of California Press.

Becker, G., A. Butler, and R. D. Nachtigall. 2005. "Resemblance Talk: A Challenge for Parents Whose Children were Conceived with Donor Gametes in the US." *Social Science & Medicine*, 61 (6): 1300–1309. https://doi.org/10.1016/j.socscimed.2005.01.018

Better Outcomes Registry and Network Ontario. 2016. *Canadian Assisted Reproductive Technologies Register Plus (CARTR Plus)*. Toronto, ON: Better Outcomes Registry and Network Ontario.

Blyth, E. 2010. "Fertility Patients' Experiences of Cross-Border Reproductive Care." *Fertility and Sterility*, 94 (1): 11–15.

Carsley, S. 2016. "DNA, Donor Offspring and Derivative Citizenship: Redefining Parentage Under the Citizenship Act." *Dalhousie Law Journal*, 39 (2): 525–559.

Cattapan, A. 2016. "Precarious Labour: On Egg Donation as Work." *Studies in Political Economy*, 97 (3): 234–252.

Faircloth, C., and Z. Gurtin. 2017. "Fertile Connections: Thinking Across Assisted Reproductive Technologies and Parenting Culture Studies." *Sociology*, 1–18. https://doi.org/10.1177/0038038517696219

Firestone, S. 1970. *The Dialectic of Sex: The Case for the Feminist Revolution*. London, UK: Johnathan Cape Ltd.

Foss, S., and W. Waters. 2011. Coding Qualitative Data. *Accessed* 14 *May* 2012. http://www.abdsurvivalguide.com/News/020603.htm

Franklin, S. 1997. *Embodied Progress: A Cultural Account of Assisted Conception*. London, UK: Routledge.

Gunby, J., & S. Daya. 2005. "Assisted Reproductive Technologies (ART) in Canada: 2001 Results from the Canadian ART Register." *Fertility and Sterility*, 84 (3): 590–599.

Hudson, N., L. Culley, E. Blyth, W. Norton, F. Rapport, and A. Pacey. 2011. "Cross-Border Reproductive Care: A Review of the Literature." *Reproductive Biomedicine Online*, 22 (7): 673–685. https://doi.org/10.1016/j.rbmo.2011.03.010

Konrad, M. 2005. *Nameless Relations: Anonymity, Melanesia and reproductive gift exchange between British ova donors and recipients*. New York, NY: Berghan Books.

Lutjen, P., A. Trounson, J. Leeton, J. Findlay, C. Wood, and P. Renou. 1984. "The Establishment and Maintenance of Pregnancy Using in Vitro Fertilisation and Embryo Donation in a Patient with Primary Ovarian Failure." *Nature*, 307, 174–175.

Nelkin, D., and S. Lindee. 1995. *The DNA Mystique*. Ann Arbor, MI: University of Michigan Press.

Pfeffer, N. 1987. "Artificial insemination, in-vitro fertilization and the stigma of infertility". In *Reproductive Technologies: Gender, Motherhood and Medicine*, edited by M. Stanworth, Cambridge, UK: Polity Press.

Ragone, H. 1994. *Surrogate Motherhood: Conception in the Heart*. Boulder, CO: Westview.

Rapp, R. 2000. *Testing Women, Testing the Fetus*. New York, NY: Routledge.

Richards, M., G. Pennings, and J. Appleby, eds. 2012. *Reproductive Donation: Practice, policy and Bioethics*. Cambridge, UK: Cambridge University Press.

Ryan, G., and H. R. Bernard. 2000. *Handbook of Qualitative Research*. Thousand Oaks, CA: SAGE Publications.

Sandelowski, M. 1993. *With Child in Mind: Studies of the Personal Encounter with Infertility*. Philadelphia, PA: University of Pennsylvania Press.

Schneider, D. 1984. *A critique of the study of kinship*. Ann Arbor, MI: University of Michigan Press.

Spar, D. 2006. *The Baby Business: How Money, Science, and Politics Drive the Commerce of Conception*. Boston, MA: Harvard Business School Press.

Statistics Canada. 2017. *Fertility: Fewer children, older moms* (Canadian MegaTrends). http://www.statcan.gc.ca/pub/11-630-x/11-630-x2014002-eng.htm.

Strathern, M. 1992. *Reproducing the Future: Anthropology, Kinship and the New Reproductive Technologies*. Manchester, UK: Manchester University Press.

Thompson, C. 2002. *Reconfiguring Kinship Studies* (pp. 175–202). Durham, NC: Duke University Press.

Thompson, C. 2005. *Making Parents: The Ontological Choreography of Reproductive Technologies*. Cambridge, MA: Massachusetts Institute of Technology.

Accounting for the money-made parenthood of transnational surrogacy

Ingvill Stuvøy

ABSTRACT

In the last decade, transnational surrogacy has attracted world-wide attention for making babies and pregnancies exchangeable with money. Involuntarily childless couples and individuals travel abroad and pay to have the desired child and to become parents. Acknowledging the importance of asking into the consequences of this monetization of reproduction, the author takes issue with universalistic assumptions about money and markets, and their presumed universal effects on social relations. Instead, it is argued that we need to explore how money works, and, by extension, how transnational surrogacy works out and becomes viable to people as a way to become parents. Putting together insights from economic sociology, and the assisted reproductive technology and parenting culture literature, the author employs the notion of *accounting* to grasp how people make sense of the money involved in making them parents. Based on a study involving 21 interviews with Norwegian gay and straight couples and single men and women seeking surrogacy abroad, the author explores how money is accounted for in three cases, set in three different countries; India, the United States and Canada. The analysis shows how money is accounted for in particular ways to confirm parenthood. These ways differ depending on the local context and transnational relations; ultimately making differentiated monetized parenthood. This is of significance when we try to conceptualize contemporary parenthood and how money seemingly sustains parenthood in ever more radical ways.

Introduction

Transnational surrogacy has in recent years attracted attention for how it connects babies, pregnancies and parenthood to money and market (e.g. Kroløkke and Pant 2012; Pande 2014; Rudrappa 2015; Vora 2015). While a welcome opportunity for those who desire a child of their own, the cross-border, monetized reproductive arrangement challenges conventional ideas of what 'makes parents' (cf. Thompson 2005), and of what can and cannot be bought for money and traded in the market (cf. Phillips 2013; Sandel 2012; Satz 2010). Hence, transnational surrogacy has emerged as a cause for concern; seen to exemplify broader tendencies such as commodification, commercialization and marketization (e.g. Hochschild 2011; Kroløkke, Foss, and Sandoval 2010; Whittaker and Speier 2010), in addition to outsourcing and exploitation (e.g. Twine 2015; Vora 2015).

These concerns direct attention towards important questions regarding contemporary economic and social life, and provide a strong mandate to examine transnational surrogacy and the question of what it means that money is introduced in the making of babies and parents. This paper is an attempt to contribute to such an examination. Here, I explore the specific ways money is arranged – handled and made sense of – as people, desiring to become parents, turn to transnational surrogacy as the way to achieve it.

By exploring the specific arrangements of money, I take issue with assumptions regarding money and markets and their presumed universal effects on social relations. Parallel to how the literature on assisted reproductive technologies (ARTs) has insisted on exploring the localization of globally existing biotechnologies (e.g. Inhorn 2003; Jasanoff 2011; Melhuus 2012), I explore the localization and differentiation of money. Drawing on the work of Viviana Zelizer (1994, 2011), I argue that we need to explore how people make sense of the money involved in their pathway to parenthood, and, by extension, how transnational surrogacy becomes viable to them as a way to become parents.

In the paper, I explore how money is handled and made sense of in three different cases; all involving Norwegian desiring-to-be parents, but set in three different countries – Canada, India and the United States, each with particular organizations of surrogacy.[1] Thus, emphasizing the need to pay attention to the specific arrangements of money in different local contexts and in different transnational relations, my method is comparative in kind. My data include a combination of interviews with Norwegian desiring-to-be parents and price lists, contracts, and payment schedules from surrogacy agencies or clinics whose services my interviewees have used. These visual displays of money (cf. Guyer 2009) are part of how the Norwegian couples and singles make sense of the money, while also indicative of specific ways of arranging the money in different local contexts. Accordingly, what I do here is breaking down the costs of parenthood in different transnational surrogacy arrangements.

In doing this, I join the on-going conversation on the specific institutionalized arrangements of surrogacy (e.g. Bharadwaj 2012; Jacobson 2016; Pande 2014; Ragoné 1994; Rudrappa 2015; Whittaker 2009). Into the conversation, I bring also perspectives from the field of parenting culture studies (cf. Faircloth and Gürtin 2017; Lee et al. 2014), emphasizing the cultural scripts on parenting, as a way to explore the logics that emerge as people account (in a double sense) for the money involved in making them parents. Here, I respond to the call from Charlotte Faircloth and Zeynep Gürtin (2017) to bridge the distinction between scholarship focused on, respectively, the bearing and the rearing of children, and contribute to a processual and more holistic understanding of how parents come to be parents.

In what follows, I first position my own endeavour within the broader literature on surrogacy and parenthood. Thereafter, I outline the theoretical framework of the paper, where I put the ART literature and parenting culture studies in conversation with economic sociology and anthropology. Moving on to the empirical study, I present the research and situate transnational surrogacy as a way for Norwegians to become parents. The succeeding analysis consists of three subsections, each presenting one case. The point I make in each part about, respectively, *timing*, *paying for the right thing* and *domestication of money*, is not necessarily at stake only in one case or one place. It is, however, present to different degrees in different cases and places, which indicates that the different regimes and different transnational relations give different arrangements of money and parenthood. I close off with a discussion of the idea of parenthood as an outcome of a differentiated process, and the significance of the specific ways such processes are tailored to make transnational surrogacy a viable pathway to parenthood.

Surrogacy and monetized parenthood

Turning to the question of parenthood in transnational surrogacy, I am drawing attention to a longstanding issue in the surrogacy and ART literature. While ARTs in general have challenged understandings of how children are conceived and parents are 'made' (cf. Thompson 2005), surrogacy[2] has been referred to as a reproductive arrangement where 'nothing guarantees one's status as the parent: not eggs, sperm, womb, or breast milk' (Pande 2014, 143). The involvement of others – the so-called 'third parties' (Blyth and Landau 2004), 'reproductive assisters' (Faircloth and Gürtin 2015), or, more specifically, surrogate mother[3] and donors[4] – has made the question of parenthood particularly salient in the case of surrogacy; causing doubt about who the parents are, how to define them, and whether they are parents at all.

The transnational version of surrogacy has added dimensions to this, as the desire of parenthood is pursued across borders (Deomampo 2015; Kroløkke, 2012; Kroløkke and Madsen 2014) and often on top of structural domination, global inequality and profound social distance (Arvidsson, Johnsdotter, and Essen 2015; Førde 2017; Rudrappa 2015; Vora 2015); bringing topicality to Shellee Colen's (1990) concept of 'stratified reproduction' (see e.g. Pande 2014).

Commercial surrogacy introduces additional layers to the 'problem' of parenthood and surrogacy; raising questions and concerns over commodification (e.g. Kroløkke et al. 2010; Phillips 2013; Radin 1996; Whittaker and Speier 2010) and marketization of reproductive bodies and intimate life (e.g. Hochschild 2012; Rudrappa 2015; Satz 2010). Pregnancy has become something one can commission for money, and, as emphasized by Amrita Pande (2014), has become labour in a double sense; at once the process of childbirth *and* an activity done (by women) to earn an income, even if often unrecognized as such (see also Jacobson 2016). This labour is arguably outsourced labour (Rudrappa 2010; Twine 2015; Vora 2015); from one country and part of the world to another – and from those with the money to pay to those who are willing to gestate and carry the child to term in return for money.

The economic activity implied in reproductive methods such as transnational surrogacy has made scholars discuss the conflation of reproduction and consumption (e.g. Lewin 2013; Mamo 2013); pointing to how 'individuals, frequently from the West, take up a flexible consumer position [...] to fulfil their dreams of parenthood' (Kroløkke et al. 2016, 7). The discussion recalls Marilyn Strathern's (1992) early predicaments in the wake of ARTs that parenthood was becoming more as if a matter of (consumer) choice.

However, as feminist economists such as Nancy Folbre (2008) have argued, all parents spend time and money on their children; costs that are often, in many contexts, underestimated[5]. Scholars on parenting culture have pointed to an intensification of this expenditure of time, energy and money within the last few decades; the result of an ideology of intensive mothering (Hays 1996) or intensive parenting (Faircloth 2014). Thus, one could argue that consumption and expenditure of money are already part of how 'normal' parenthood in the Western world is established and substantiated.

Commercial surrogacy seems nonetheless to exaggerate the issue of money in the making of parents. First, money – as an asset – is required for people to *become* parents, not only to *do* parenting. Second, in surrogacy, the desiring-to-be parents commission the making of the baby, and pay someone to gestate and give birth to the child for them. The commissioning distinguishes surrogacy from adoption, which patently also brings money, affluence and babies together (see e.g. Yngvesson 2002), while the gestation and birth distinguish surrogacy from other types of assisted reproduction, and arguably contribute in making the monetized exchange more apparent (Spar 2006, 70f). This exchange involves not only the desiring-to-be parents, but involves also non-parents; those who are 'birthing mothers' (or parents), as Elly Teman (2010) has conceptualized it, but not themselves becoming parents from the act of birthing. A pertinent question, I contend, is how money is handled and made sense of in an exchange and arrangement that is meant to make some people (the only) parents and others not.

Making sense of money and parenthood

Economic sociologist Viviana Zelizer has referred to surrogacy as a 'risky exchange'; often seen to be going 'beyond the boundary of decency' (Zelizer 2011, 288). Confronted with such exchanges, Zelizer emphasizes the importance of exploring how they actually work – and work out for its participants. In her broader work, Zelizer has demonstrated the shortcomings of universalistic assumptions about money and economic life, and instead emphasized the multiple and differentiated ways people relate to and handle money. Posing the concrete interweaving of social and monetary relations as her analytical question (cf. Fourcade 2012, 1057), Zelizer explores how money is arranged by the participants through 'earmarking' money, making it into *monies* (Zelizer 1994), and through 'relational work' that seeks to create 'good matches' between meaningful relations, transactions and money (Zelizer 2011).

Zelizer's attention to 'relational work' and 'good matches' has affinities with ART scholar Charis Thompson's (2005) widely disseminated idea of 'ontological choreography', where things of different ontological order – such as intentions, sperm, ova, and, also, money – are strategically choreographed in the fertility clinic to confirm parenthood and 'make parents'. The idea that parenthood is something made rather than given finds, moreover, an equivalent in the emphasis within the parenting culture literature on 'parenting' as a cultural accomplishment and product (e.g. Faircloth, Hoffman, and Layne 2013). Putting these literatures together provides a way to examine the specific ways money is arranged to confirm and disambiguate parenthood.

The literature on parenting culture emphasizes the importance put on reflexivity among parents. Parents are required to account for and ensure that their parenting is 'good enough'; that is, in accordance with cultural scripts providing rules on how to parent (Faircloth and Gürtin 2017, 5).[6] In the context of ARTs, requirements to account for one's parenting 'extends backwards' (Faircloth and Gürtin 2017, 8), as people are expected to account for their parental commitment even before becoming parents. Additionally, in the case of the desiring-to-be parents in transnational surrogacy, accounting could be seen as having a double meaning; referring to their parental claim and to the money involved in their pathway to parenthood.

While Zelizer's idea of relational work mainly captures how such accounting is done at an interactional level, her work could also be seen as pointing towards an institutional order as people are 'working through institutions (and sometimes against them)' (Fourcade 2012, 1060) to maintain meaningful relations. Within the surrogacy and ART literature, attention has been devoted to the institutional arrangements that enable the marketization of egg, sperm and pregnancy (e.g. Almeling 2011; Rudrappa 2015).

In transnational surrogacy, the institutional is at stake on several levels: desiring-to-be parents relate to state regulations on parenthood, citizenships and markets (Deomampo 2015); they interact with agencies and clinics (Rudrappa and Collins 2015), and they relate to – and count – money as they are presented to them in price lists, invoices and contracts. Drawing on a paper by economic anthropologist Jane Guyer (2009) on the composition of price – and on how price is produced, presented, revealed and concealed – I will in the analysis pay attention to the material arrangements of money as it becomes visible in price lists and contracts. This, I contend, enables exploration of particular institutionalized arrangements of money and how these come to influence how money is seen and, by extension, understood by those whose parenthood is at stake and in the making.

The research

In this paper, the people whose parenthood I am interested in are Norwegian involuntarily childless couples and individuals, who have travelled abroad to have a surrogate mother carry and give birth to their children. I have conducted 21 interviews with these couples and individuals, by now all parents. Eleven of these interviews were with gay men; most of them being in a couple at the time of seeking out surrogacy, apart from one single man. Ten interviews were conducted with straight men and women. Of these, one was a single woman. Of the total 21 interviews, 10 were conducted with couples, while 11 were with individuals. Most of my interviewees had travelled or were to travel to the United States.; three couples and one single man had travelled to India[7]; while three couples had travelled to Canada.[8]

The costs of transnational surrogacy are significant for who these desiring-to-be parents are in terms of age, occupation and class background. My interviewees were between the ages of 28 and 50, and had generally high levels of education and an annual income above or at average level in Norway, which in 2016 was about $62,200 US dollars. All the interviews were done between February 2013 and December 2014. I carried out the interviews quoted here in Norwegian and translated the quotations into English; all names and other identifying details have been changed.

Situating transnational surrogacy in Norway and beyond

Transnational surrogacy 'arrived' in Norway around 2010 following the travels taken by Norwegian involuntarily childless couples and individuals to have someone else, in a different country and part of the world, gestate and give birth to their children (Andersen 2013; Stuvøy 2016). This reproductive travelling caused criticism of 'baby shopping' and 'human trafficking', and a commentator in Norway's largest newspaper wrote: '*People sit in front of their computers and shop new lives. Latest fashion: Your own biological child. Before 1 million [NOK], now only 30,000 [NOK]*' (Moen 2010). Accentuating the numbers at stake – even if not being correct numbers: 30,000 should most likely have been 300,000 – the commentator drew attention to money's involvement in these reproductive travels taken in a quest for parenthood.

While surrogacy is controversial and not legal in Norway,[9] travelling abroad for surrogacy is not illegal for Norwegian citizens. Norwegian authorities grant citizenship to the children – born by a surrogate mother abroad – through the Norwegian genetic father[10]; making it feasible to travel abroad to countries where surrogacy is legal and accessible to foreigners.[11] The most frequented destinations for surrogacy among Norwegians have been the United States and India. While the surrogacy traffic from Norway to India was central in bringing surrogacy on the agenda in Norway, this traffic effectively stopped in 2013 when the Indian authorities introduced new visa requirements for foreigners using surrogacy in the country.[12] Today, the United States seems to be the most popular destination for surrogacy among Norwegians (Dommerud and Tjernshaugen 2017). Canada is another destination present in my material, although less frequented. Norwegians are also known to be travelling to some other countries, such as Georgia and Ukraine.

Citizenship for the child is of significance for transnational surrogacy to work out for the desiring-to-be parents (cf. Deomampo 2015). The failure to establish citizenship for children born to a Norwegian single woman by an Indian surrogate mother was in fact what brought surrogacy to the public's attention in Norway (see Kroløkke, 2012). In both India and Norway, citizenship follows from the parents; meaning that parenthood needs to be established for the child to attain citizenship. This is different in the United States and Canada, where the child acquires citizenship by virtue of being born there.

In the United States[13] and India, surrogacy has been organized as a commercial arrangement, where the surrogate mother is paid an agreed-upon amount. In Canada, surrogacy is non-commercial or altruistic. Any payment apart from expenses to the surrogate mother is not allowed (Health Canada 2013). Among my interviewees, the costs of surrogacy have varied between around $35,000 and $400,000 US dollars; depending on country of destination and the number of children, pregnancies and attempts. While the United States is the more expensive destination for surrogacy, India has been among the cheaper. My interviewees travelling to the United States spent about $120,000 US dollars on surrogacy, while those travelling to India spent about $35,000–40,000 US dollars. Canada is, in terms of price, between the two; the couples I have interviewed have spent about $75,000 US dollars. Thus, it is clear that for my Norwegian interviewees, money has been involved in all their endeavours to become parents, regardless of whether surrogacy is organized as commercial or non-commercial.

Becoming (Norwegian) parents

The question that guides me throughout the analysis is how money is accounted for: how money is arranged, handled and made sense of to confirm the parenthood of the Norwegian desiring-to-be parents, and, by extension, make transnational surrogacy work out for them. I have chosen to feature three cases with Norwegians travelling to, respectively, India, the United States and Canada. These cases are chosen for how they each illustrate a particular feature of how money is arranged. Neither case exhausts the ways of monetization of parenthood in each setting, but illustrates instead different ways this takes place. The cases do not distinguish themselves in prominent ways from the others among my interviewees that have travelled to the same country. My aim here is to portray the logics at stake, as these become attainable through the parents' narratives and the material representations of the money.

Timing is everything: Øyvind and Morten in India

I start with the arrangement of money in Øyvind and Morten's story of becoming parents through surrogacy India. In their story, the timing of the money emerged as significant to how their parenthood was established and affirmed.

Øyvind and Morten are married fathers of two children, who were born a few months apart by two different Indian surrogate mothers. In our interview, Øyvind and Morten told me about their very first payment: made with a VISA card on their first trip to India. The payment was for the freezing of their sperm. The two had travelled to India to check out the clinic they had been recommended. After agreeing between them that the clinic seemed to satisfy their expectations, Øyvind and Morten decided to get the process started. The freezing of the sperm and the money to pay for it, marked to Øyvind and Morten that they were in the process of becoming fathers.

The two men then went back to Norway, from where they transferred the rest of the money in stages throughout the process. In the price list from their clinic in New Delhi, the costs of surrogacy were listed in stages: 'Stage One', 'Stage two' and 'Stage three', consisting of a total of six instalments. Noteworthy, in this price list, the baby is not listed as an item. This absence supports the idea of surrogacy as not about purchasing babies. In Øyvind and Morten's understanding, the children were not the result of a purchase – and as genetic fathers, they had not purchased the children. They were, as Morten put it, '*just as much a dad as any other dads*'.

Discussing the issue of motherhood, Øyvind and Morten referred to the two surrogate mothers, the egg donor, and the female doctor as four women, who all contributed to the process with different services. The split motherhood ensured, in Øyvind and Morten's experience, that none of these women could claim to be the mother of the children. Thus, Øyvind and Morten strategically choreographed (cf. Thompson 2005) the different items necessary in the baby-making process in accordance to a culturally familiar script of the significance of the gene in the Western world (see e.g. Nelkin and Lindee 2004). Accentuating their sperm above other items and other potential parents, Øyvind and Morten confirmed themselves to be *the* parents.

In this process, the order of things, including money and invoices, was of significance. Showing me their correspondence with the clinic in Delhi, Øyvind and Morten credited the clinic for keeping things apart. They explained to me how they first got notified by one division of the clinic that a heartbeat, and pregnancy, was confirmed, and then shortly after got an e-mail from the economy division; notifying them that as heartbeat was confirmed, a first instalment to the surrogate mother was now due.

It seems significant that the money from Øyvind and Morten to the surrogate mother is due at confirmation of heartbeat. To listen for a heartbeat is a common way among medical personnel – not only in the surrogacy clinic, but more generally – to confirm life underway. As such, the centrality of the heartbeat is not necessarily that astounding. Yet, it attracts my attention for the way it explicitly connected money and body, and used one – the body – to set the other – the money – in motion. Correspondingly, the money stops if the heartbeat stops: In the case of pregnancy loss, the money from Øyvind and Morten to the surrogate mother would continue for one month and then cease. Other costs would cease immediately, apart from professional costs for post-natal care of surrogate mother.

The heartbeat – although located within the body of the surrogate mother – seems thus to point towards the baby in the making (see Rudrappa 2015, 126ff). Linking the money closely to the future baby, one could read this as an illustration of how the money is arranged to draw the baby closer to Øyvind and Morten – and confirm the surrogacy process as *their* process of becoming fathers.

The arrangement of the money contributed, moreover, in making the surrogate mothers and egg donors 'prosthetic' (Thompson 2005, 145). The money moved from Øyvind and Morten to the clinic, who then paid the surrogate mothers and the egg donor for their 'services', as Øyvind and Morten referred to it. The two only dealt with the clinic and never met the egg donor; reflecting the institutional set-up commonly prescribed by Indian surrogacy clinics (Førde 2017, 162; Rudrappa 2015, 136). Øyvind and Morten only met each of their surrogate mothers once during the whole process; on both occasions at the Norwegian embassy in New Delhi, as a requirement from Norwegian authorities to issue passports to children born through surrogacy. The lack of contact with the surrogate mothers and egg donor kept these women at a distance to the process (see Rudrappa 2015, 2016), as it was experienced by Øyvind and Morten, who emerged as the only *parents* involved.

Paying for the right thing: Monika in the United States

Timing is also important for those among my interviewees travelling to the United States. However, as the case of Monika indicates, parenthood is established by paying not only at a particular time, but also by paying for the *right* thing. This points, for one, towards the importance of keeping things apart within the concrete arrangement, but was also a way to set her own way of doing surrogacy apart from that of others, and in particular the way it was done in India.

Monika was a married mother of one and expecting twins carried by an American surrogate mother at the time of our interview, where her husband was not present. Monika told me that she and her husband had, at one point, joked about going to the United States to '*buy a baby*'. This was before they seriously considered surrogacy, and while the thought of it was still very new and somewhat strange to them. By the time of our interview, Monika was insisting to me that the baby was not at all what they were paying for.

> [My husband] quickly understood that what we were buying was a medical service. That's what the money went to. [...] And then I understood that it's not the child we're paying for. We're paying for the medical [parts], and, also, that they [the agency] are organizing everything for us. Documents; contracts; they find the surrogate mother; the matching. [...] We're paying for the process. Not one krone is changing hands for the baby. Not one krone.[14]

Monika emphasized it was medical services and the *process* they were paying for, and not the baby. The idea that what is bought and paid for are services – and a whole bundle of them – can be re-found in the price lists and invoices distributed by the different American surrogacy agencies used by the Norwegian desiring-to-be parents. One American agency presented the different costs over three full pages. Another American agency, the one used by Monika and her husband, operated with eight main headlines in their price list, and in total more than 40 items were listed as what composed the costs of the surrogacy arrangement. As visual arrangements of the money, the detailed breakdown of costs suggests that the desiring-to-be parents pay for a bundle of services – and also that they pay a bundle of different people, all playing some part in the surrogacy process. The surrogate mother, then, was not the only one contributing, but rather one out of many. Thus, this arrangement diminished the importance of the role played by any one contributor.

Again, it is notable that at no point do the price list mention the baby. The baby is, as Monika emphasized, not what is bought and paid for. This is also explicitly stated in the contract; a central document to ensure that the right thing is paid for. I am citing here a California sample contract, used by Monica's agency; an agency popular among my Norwegian interviewees. The contract says: *'It is expressly understood that this Agreement in no way constitutes payment for a child or relinquishment of a child, or payment for consent to adoption.'* The money involved in the surrogacy arrangement is thus not to be understood as a sum that represents a price for a baby, where the baby is an object for sale, or as payments to people for them to relinquish their child. Rather, these payments should be understood as *'reimbursements'* to the one carrying the child for the, by the contract, already designated *'intended parents'*. The contract defines not only parenthood, but also the money transferred to and from the signing parties. This particular earmarking (Zelizer 1994) of the money marks a distinction made between the money to the surrogate mother and the money going to others.[15]

Notably, the reimbursement to the surrogate mother is for pain and suffering, and not for services. The American surrogate mother is neither selling the baby nor doing a job, making her labour disappear from view (see e.g. Jacobson 2016). Instead, the surrogate mother is positioned as someone to whom the Norwegian parents relate on a personal level; making the exchange less about money and market. To Monika, this also distinguished her own surrogacy journey from one undertaken in India, which at the time was much debated in Norway. Emphasizing that India was never an alternative destination for her, in part for the lack of contact to the surrogate mother, Monika reflexively set apart her way of doing transnational surrogacy from the Indian arrangement. Doing this, Monika seemed to account as much for her moral upper hand as for the money.

The personal relationship was something stressed and facilitated for by the American surrogacy agencies. Referring to the relationship as *'the backbone of surrogacy'*, a director of one of the American surrogacy agencies told me that they made sure that their clients should not have to talk money with their surrogate mother, and said: *'We take the money out of the equation'.* Money, then, was something the desiring-to-be parents should not have to account for as part of their relation to the surrogate mother. This was part of what made the surrogacy arrangement work out for Monika and her husband; confirming not only that money was paid correctly and for the right things, but also that money was not defining their relation to the woman birthing their child.

Bringing the money closer to home: Line and Ole in Canada

I met Line and Ole at their home in Norway, with the twins – born by a Canadian surrogate mother – running around, already a few years old and talking by the time of our interview. Characteristic of Line and Ole's narrative about the money involved in making them parents was what I refer to as a *domestication* of the money. The money appeared as if comparable to money at home; making the process appear similar to a reproductive process in Norway – and their parenthood similar to that of other Norwegians, in particular those in need of assistance.

In their account of the money, Line and Ole emphasized that the money spent on surrogacy was mainly money spent on medical services. As such, their understanding of the money resembled Monika's description above. Nonetheless, the point I would like to make here is that the emphasis on medical services is not only about paying for the right thing, but also about making the money less about a market, and more about health. Health is, for Norwegians, something that is paid collectively through taxes and accessed through the welfare state. Fertility services, more specifically, are subsidized by the state, with smaller user fees.

In that regard, the existence of a public health system in Canada was among the characteristics emphasized by the couples who travelled there for surrogacy, including Line and Ole. Line and Ole referred to '*structural and cultural similarities*' between Norway and Canada as part of what made Canada their preferred destination, referring explicitly to the two countries' health systems. Money in Canada, as they saw it, had more to do with money spent on state-subsidized welfare services than it had to do with a full-blown commercial market.

This was also reflected in the understanding of the more moderate costs of surrogacy in Canada. Overall in my material, the differences in costs are often referred to as a reason why my interviewees preferred one country over another. In the case of Canada, the relative lower costs compared to the United States was a reason to go there for the parents I interviewed. While this was a question of what they could afford, it was also posed as a question of the degree of commercialization and marketization of surrogacy in the two countries. This was epitomized in the question of payment to the surrogate mother. To the extent surrogacy has been discussed as something that could or ought to be legal in Norway, it has almost without exception been altruistic surrogacy that has been the case. To Ole and Line, Canada was their preferred pick due precisely to the altruistic way of organizing surrogacy. Instead of money, Line and Ole gave gifts to the surrogate mother and her children, and wrote a letter expressing their gratitude.

Explicating to me how they made the choice of Canada, they referred to India as a contrast; as how they would not like to do surrogacy. In their understanding, surrogacy in India seemed more as a transaction in a market, where pregnancy was done for money, rather than the '*help*' they received from their Canadian surrogate mother. However, in our interview, Line and Ole conceded to having changed their mind somewhat with regard to the money to the surrogate mother, having become more '*liberal*' on this point, as Ole referred to it. While it was important to them at the time when they chose Canada, by the time of our interview Line and Ole thought it would be unproblematic to pay the surrogate mother.[16]

Not paying the surrogate mother for what she did for them, Line and Ole described the surrogate mothers as someone '*babysitting*' their children for nine months. Thus, as in the case of Øyvind and Morten, the surrogate mother appeared as someone providing a type of service. Babysitting invokes an idea of odd jobs done for instance by a neighbour's older child, in return for some pocket money. Childcare is, however, also paid labour in contemporary Norway, though widely associated with the public sector and the welfare state. Conceiving of what the surrogate mother did as babysitting could be understood as a way of bringing them, as parents, closer to the 'normal' Norwegian parent, who leaves his or her children in day care facilities, which are welfare services rather than services in a market.

Finally, Ole stressed that another large budget item was the travelling to and from Canada. While medical expenses would have been covered by the state if surrogacy could be done in Norway, travel expenses would have been altogether unnecessary. The money spent, then, was the price to pay to go abroad. This makes their parenthood comparable to the parenthood of other Norwegians; the only difference being that they had to travel abroad to become parents.

Conclusion: parenthood as (differentiated) outcome

Put together, these cases point to how money is accounted for – arranged, handled and made sense of – differently in different local contexts and different transnational arrangements. It illustrates that money is differentiated and locally situated, doing different things depending on where and between whom it is introduced, even if also global in character and travelling across borders. Additionally, it indicates that different transnational arrangements make different kinds of monetized parenthood. The 'good matches' (cf. Zelizer 2011), affirming and establishing parenthood, differ depending on where people travel for transnational surrogacy.

Yet, while money is differentiated in these transnational surrogacy arrangements involving Norwegian desiring-to-be parents, there are also similarities in how money and parenthood come together. One striking similarity is how transnational surrogacy seemed to work out for all my interviewees, regardless of where they travelled. All of them are now living in Norway with the children born to them by a surrogate mother in either India, the United States or Canada. As such, the money worked to similar outcomes, even if they were handled and made sense of differently.

This points towards a shared idea among my interviewees of parenthood as the result of a *process*. By paying for the process rather than the end result – the baby – they become parents. The process is what makes it possible to transfer motherhood from the woman conventionally thought to be the mother – the one giving birth – to another person. This makes my Norwegian interviewees, as the ones commissioning the process and intending to parent, *the* parents. The idea of parenthood as the result of a process coincides with Thompson's (2005) description of the making of parents in the American fertility clinic. It also recalls Strathern's idea that with surrogacy we are moving away from the 'real' towards that which works and has effect: 'The mother is simply she who has been made effective' (Strathern 1998, 202). Strathern adds, however, that the *effective* parent 'may have to sustain that judgement by continuing to give evidence to the capacity to parent' (Strathern 1998, 202); echoing the insights from parenting culture studies that parenting requires continuous reflection on whether one is a good (enough) parent (Faircloth and Gürtin 2017, 5).

Notably, several of my interviewees would vehemently argue that their process is nothing alike the process of others – and in particular, nothing alike the process of surrogacy in India. While my foregoing analysis does show there are differences between the processes, distinction (cf. Bourdieu 1984) should also be considered as part of this differentiation of the processes in different countries. As Petersen, Kroløkke, and Myong (2017) have argued with regard to the surrogacy debate in Norway, 'stances for or against surrogacy are reworked through the distinction between ethical and unethical uses of surrogacy' (105). This points towards how the work done to find good matches and make good parents implies also the construction of bad matches and perhaps even bad parents.

In her review of Zelizer's work, economic sociologist Marion Fourcade (2012) points to Zelizer's silence about 'bad matches'. Fourcade asks into 'the 'institutional order' that keeps people in line behind the 'interaction order'' (Fourcade 2012, 1059), and comments that the work people do to make things work out is 'a powerful reminder that individuals differentiate because they (and their personal relations) are themselves differentiated socially'' (Fourcade 2012, 1059). In transnational surrogacy, this is the case on multiple levels. Different surrogacy journeys and destinations for surrogacy are distinguished from one another; and the reproductive assistance provided is priced differently depending on who performs it and where it is performed. The distinction made between parents and non-parents is, moreover, made possible within an institutional order where stratification is blatantly present. Understanding how 'good matches' are made provides, nonetheless, knowledge about the logics – and ways of accounting – that make it possible for people to enter into a stratified and monetized reproductive terrain, and leave as parents.

Notes

1. In India, surrogacy has since August 2016 not been legal for foreigners, after the Indian government presented a new draft law banning commercial surrogacy and prohibiting surrogacy for foreign nationals.
2. Surrogacy can be defined as a reproductive arrangement where a woman gestates and gives birth to a child on behalf of someone else, who are to parent the child. A distinction can be made here between *traditional* and *gestational* surrogacy. In the case of the former, the woman gestating and giving birth also contributes with the eggs used to conceive the child, making the child her genetic offspring. In gestational surrogacy, in contrast, the child is conceived with eggs from either the desiring-to-be mother or a so called egg donor (see note 4). My interviewees have all used gestational surrogacy, either using their own eggs or purchasing eggs from a donor.
3. I would write surrogate mother here to note the ambivalence concerning this woman and her reproductive contribution, and the problem of finding descriptive and politically responsible concepts in a conflicted terrain. Inspired by Derrida's (2016, orig. 1974) elaboration on the heideggerian idea of *sous rature*, I cross out 'mother' since the term 'surrogate mother' is inaccurate and potentially problematic for the gendered ideas it conveys about the reproductive labour performed (see e.g. Pande 2014). Yet, as the term 'surrogate' – or other common alternatives such as 'gestational carrier' – is no more accurate or politically responsible, the word 'mother' seems to be necessary and remains therefore legible. For an insightful discussion of

the terms used to denote the woman gestating and birthing the children in surrogacy arrangements, see Bharadwaj (2012). Throughout the paper, I will write 'surrogate mother', which is the most established term in Norwegian. As I see it, 'surrogate mother' also does more work to sustain ambiguity in a paper on how parenthood is made in transnational surrogacy.

4. The terms donor and donation reflect a language of the gift, illustrating how the market in eggs and sperm is discursively formatted (see Almeling 2011; Gupta 2006).
5. Folbre writes predominantly about the American context, which on the issue of public spending on child care is quite distinct from the Norwegian welfare state context. See e.g. Ellingsæter and Pedersen (2013).
6. Faircloth and Gürtin (2017) emphasize that the cultural script requiring intensive parenting is intimately associated with the middle-class in a Euro-American context. They argue, nonetheless, that this script is increasingly something people world over relate to and take into account in their own parenting.
7. The traffic of Norwegian desiring-to-be parents to India stopped around the time I started doing interviews, making recruitment harder.
8. While a low number of interviewees, it might nonetheless be an overrepresentation compared to the other two destinations present in my material. The Norwegian Embassy in Canada has registered only six children born through surrogacy in Canada in the period between 2008 and November 2016. Although not everyone registers with the embassy, this indicates that the numbers are low.
9. Importantly, surrogacy is not explicitly illegal in Norway. Instead, so called gestational surrogacy is illegal by implication due to a ban on egg donation, dictating that eggs removed from a woman must be returned to the same woman (The Biotechnology Act 2003, 2–15). Moreover, contracts on surrogate motherhood are declared void (The Children's Act 1981).
10. Motherhood is defined in Norwegian law as the woman giving birth. In the context of surrogacy, this means the surrogate mother is always the mother. Norwegian authorities do however acknowledge the desiring-to-be fathers' parenthood. Fatherhood is recognized based on declaration if the surrogate mother is unmarried, or declaration and DNA test in the cases where the surrogate mother is married.
11. This distinguishes Norway from other European countries, such as France, Switzerland, Italy and Germany, where the authorities have been unwilling to recognize the parenthood of those travelling abroad for surrogacy.
12. According to the Norwegian Embassy in New Delhi, the numbers had been steadily going up since the embassy registered their first two cases in 2008, reaching a record high 34 babies born to Norwegians by Indian surrogate mothers in 2012. The numbers dropped to 19 in 2013, only 1 in 2014, and 0 in 2015.
13. Surrogacy law is defined at state level. Not all American states allow surrogacy. California has been a main destination for surrogacy among Norwegians, but also other states on the East coast have become popular in recent years for their relative proximity in terms of geographical distance to Norway.
14. Krone (NOK) is the currency in Norway. There are about 8 kroner in $ 1 US dollar.
15. The particular earmarking of the money made visible in the contract could be interpreted as intimately linked to the legacy of slavery in the United States (e.g. Thompson 2005, 145).
16. One of the couples I interviewed did in fact pay their Canadian surrogate mother – in cash – upon the birth of their child; describing it as a type of return-gift from them to her.

Ethical approval

The research project has been notified to and ethically reviewed by the Data Protection Official for Research, sorting under the Norwegian Centre for Research Data (NSD). NSD is responsible for ensuring that research abides to the Norwegian Personal Data Act and Health Register Act.

Acknowledgments

I would like to thank the anonymous reviewers for very helpful and encouraging comments, and also extend this thank you to the editors of this special issue, Charlotte Faircloth and Zeynep Gürtin, and to my supervisors Arve Hjelseth and Charis Thompson. I would also like to thank Elise Farstad Djupedal for reading and commenting on different drafts of this paper.

Disclosure statement

No potential conflict of interest was reported by the author.

Funding

The research has been funded by the Norwegian University of Science and Technology (NTNU).

References

Almeling, R. 2011. *Sex Cells: The Medical Market for Eggs and Sperm*. Berkeley, Los Angeles, and London: University of California Press.

Andersen, U. C. 2013. "Offerposisjonens paradoks – Offentlig debatt om surrogati." *Tidsskrift for samfunnsforskning* 54 (1): 31–62.

Arvidsson, A., S. Johnsdotter, and B. Essen. 2015. "Views of Swedish Commissioning Parents Relating to the Exploitation Discourse in Using Transnational Surrogacy." *Plos One* 10 (5): 1–12.

Bharadwaj, A. 2012. "The Other Mother: Supplementary Wombs and the Surrogate State in India." In *Reproductive Technologies as Global Form. Ethnographies of Knowledge, Practices, and Transnational Encounters*, edited by M. Knecht, M. Klotz, and S. Beck, 139–160. Frankfurt and New York: Campus Verlag.

Blyth, E., and R. Landau. 2004. *Third Party Assisted Conception across Cultures: Social, Legal and Ethical Perspectives*. London and New York: Jessica Kingsley Publishers.

Bourdieu, P. 1984. *Distinction: A Social Critique of the Judgement of Taste*. Cambridge: Harvard University Press.

Colen, S. 1990. "Housekeeping for the Green Card: West Indian Household Workers, the State, and Stratified Reproduction in New York." *At Work in Homes* 3: 89–118.

Deomampo, D. 2015. "Defining Parents, Making Citizens: Nationality and Citizenship in Transnational Surrogacy." *Medical Anthropology* 34 (3): 210–225.

Derrida, J. 2016. *Of Grammatology*. Translated and edited by G. C. Spivak. Baltimore, MD: John Hopkins University Press.

Dommerud, T., and K. Tjernshaugen. 2017. "Populært og dyrt med surrogati-barn fra USA." *Aftenposten*, February 13.

Ellingsæter, A. L., and E. Pedersen. 2013. "Fruktbarhetens fundament i den norske velferdsstaten." *Tidsskrift for samfunnsforskning* 54 (1): 3–29.

Faircloth, C. 2014. "Intensive Parenting and the Expansion of Parenting." In *Parenting Culture Studies*, edited by E. Lee, J. Bristow, C. Faircloth, and J. Macvarish, 25–50. London: Palgrave Macmillan.

Faircloth, C., and Z. Gürtin. 2015. "Making Parents: Assisted Reproduction and Contemporary Parenting Culture." Paper presented at the annual meeting of the American Anthropology Association, November 17–22.

Faircloth, C., and Z. B. Gürtin. 2017. "Fertile Connections: Thinking across Assisted Reproductive Technologies and Parenting Culture Studies." *Sociology* 1–18.

Faircloth, C., D. M. Hoffman, and L. L. Layne. 2013. *Parenting in Global Perspective: Negotiating Ideologies of Kinship, Self and Politics*. London and New York: Routledge.

Folbre, N. 2008. *Valuing Children: Rethinking the Economics of the Family and Public Policy*. Cambridge, MA: Harvard University Press.
Fourcade, M. 2012. "The Moral Sociology of Viviana Zelizer." *Sociological Forum* 27 (4): 1055–1061.
Førde, K. E. 2017. *Intimate Distance. Transnational Commercial Surrogacy in India*. PhD diss., University of Oslo, Oslo.
Gupta, J. A. 2006. "Towards Transnational Feminisms. Some Reflections and Concerns in Relation to the Globalization of Reproductive Technologies." *European Journal of Women's Studies* 13 (1): 23–38.
Guyer, J. I. 2009. "Composites, Fictions, and Risk: Toward an Ethnography of Price." In *Market and Society: The Great Transformation Today*, edited by C. Hann and K. Hart, 203–220. New York: Cambridge University Press.
Hays, S. 1996. *The Cultural Contradictions of Motherhood*. New Haven, CT: Yale University Press.
Health Canada. 2013. "Prohibitions Related to Surrogacy." http://www.hc-sc.gc.ca/dhp-mps/brgtherap/legislation/reprod/surrogacy-substitution-eng.php
Hochschild, A. R. 2011. "Emotional Life on the Market Frontier." *Annual Review of Sociology* 37: 21–33.
Hochschild, A. R. 2012. *The Outsourced Self. Intimate Life in Market Times*. New York: Metropolitan Books, Henry Holt and Company.
Inhorn, M. C. 2003. *Local Babies, Global Science: Gender, Religion, and In Vitro Fertilization in Egypt*. New York and London: Routledge.
Jacobson, H. 2016. *Labor of Love: Gestational Surrogacy and the Work of Making Babies*. New Brunswick, New Jersey, and London: Rutgers University Press.
Jasanoff, S. 2011. *Designs on Nature: Science and Democracy in Europe and the United States*. Princeton, NJ: Princeton University Press.
Kroløkke, C. 2012. "From India with Love: Troublesome Citizens of Fertility Travel." *Cultural Politics an International Journal* 8 (2): 307–325.
Kroløkke, C., K. A. Foss, and J. Sandoval. 2010. "The Commodification of Motherhood: Surrogacy as a Matter of Choice." In *Contemplating Maternity in an Era of Choice: Explorations into Discourses of Reproduction*, edited by S. Hayden and D. L. O. B. Hallstein, 95–114. Plymouth: Lexington Books.
Kroløkke, C., and K. H. Madsen. 2014. "Moderskab (elser)." *Kvinder, Køn and Forskning* 23 (1/2): 70–81.
Kroløkke, C., L. Myong, S. W. Adrian, and T. Tjørnhøj-Thomsen. 2016. *Critical Kinship Studies*. London and New York: Rowman & Littlefield International.
Kroløkke, C., and S. Pant. 2012. "'I Only Need Her Uterus': Neo-Liberal Discourses on Transnational Surrogacy." *NORA* 20 (4): 233–248.
Lee, E., J. Bristow, C. Faircloth, and J. Macvarish, eds. 2014. *Parenting Culture Studies*. London: Palgrave Macmillan.
Lewin, E. 2013. "Embracing Consumption: Making Sense of Gay Fathers' Strategies for Becoming Parents." Paper presented at the Austin Summit on LGBT Families, University of Texas.
Mamo, L. 2013. "Queering the Fertility Clinic." *Journal of Medical Humanities* 34 (2): 227–239.
Melhuus, M. 2012. *Problems of Conception: Issues of Law, Biotechnology, Individuals and Kinship*. New York and Oxford: Berghahn Books.
Moen, E. S. 2010. "Designbaby - nå bare kr. 30.000. [Design Baby – Now Only 30,000 NOK]." *VG*, March 18.
Nelkin, D., and M. S. Lindee. 2004. *The DNA Mystique: The Gene as Cultural Icon*. Ann Arbor, MI: University of Michigan Press.
Pande, A. 2014. *Wombs in Labor: Transnational Commercial Surrogacy in India*. New York: Columbia University Press.
Petersen, M. N., C. Kroløkke, and L. Myong. 2017. "Dad & Daddy Assemblage: Resuturing the Nation through Transnational Surrogacy, Homosexuality, and Norwegian Exceptionalism." *GLQ* 23 (1): 83–112.
Phillips, A. 2013. *Our Bodies, Whose Property?* Princeton, NJ; Oxford: Princeton University Press.

Radin, M. J. 1996. *Contested Commodities: The Trouble with Trade in Sex, Children, Body Parts, and Other Things*. London: Cambridge University Press.
Ragoné, H. 1994. *Surrogate Motherhood. Conception in the Heart*. Boulder, CO; San Francisco, CA; Oxford: Westview Press.
Rudrappa, S. 2010. "Making India the 'Mother Destination': Outsourcing Labor to Indian Surrogates." *Research in the Sociology of Work* 20: 253–285.
Rudrappa, S. 2015. *Discounted Life: The Price of Global Surrogacy in India*. New York and London: New York University Press.
Rudrappa, S. 2016. "Conceiving Fatherhood: Gay Men and Indian Surrogate Mothers." In *Globalized Fatherhood*, edited by M. C. Inhorn, W. Chavkin, and J.-A. Navarro. New York and Oxford: Berghahn Books.
Rudrappa, S., and C. Collins. 2015. "Altruistic Agencies and Compassionate Consumers: Moral Framing of Transnational Surrogacy." *Gender & Society* 29 (6): 937–959.
Sandel, M. J. 2012. *What Money Can't Buy: The Moral Limits of Markets*. New York: Farrar, Straus and Giroux.
Satz, D. 2010. *Why Some Things should not be for Sale: The Moral Limits of Markets*. Oxford and New York: Oxford University Press.
Spar, D. L. 2006. *The Baby Business: How Money, Science, and Politics Drive the Commerce of Conception*. Boston: Harvard Business Press.
Strathern, M. 1992. *Reproducing the Future: Essays on Anthropology, Kinship and the new Reproductive Technologies*. Manchester: Manchester University Press.
Strathern, M. 1998. "Surrogates and Substitutes: New Practices for Old?" In *The Politics of Postmodernity*, edited by J. Good and I. Velody, 182–209. Cambridge: Cambridge University Press.
Stuvøy, I. 2016. "Surrogati: Reproduksjon i globalt farvann" In *Reproduksjon, kjønn og likestilling i dagens Norge*, edited by M. N. Ravn, G. K. Kristensen, and S. Ø. Sørensen, 195–221. Oslo: Fagbokforlaget.
Teman, E. 2010. *Birthing a Mother: The Surrogate Body and the Pregnant Self*. Berkeley, Los Angeles, and London: University of California Press.
The Biotechnology Act. 2003. "Lov-2003-12-05-100 Lov om humenmedisinsk bruk av bioteknologi m.m. [Act Relating to the Application of Biotechnology in Human Medicine, etc.]." https://www.regjeringen.no/en/dep/hod/id421/
The Children's Act. 1981. "[LOV-1981-04-08-7 Lov om barn og foreldre (barnelova)] [Act No. 7 of 8 April 1981 Relating to Children and Parents]." https://lovdata.no/dokument/NL/lov/1981-04-08-7
Thompson, C. 2005. *Making Parents: The Ontological Choreography of Reproductive Technologies*. Cambridge and London: MIT Press.
Twine, F. W. 2015. *Outsourcing the Womb: Race, Class and Gestational Surrogacy in a Global Market*. New York and London: Routledge.
Vora, K. 2015. *Life Support: Biocapital and the new History of Outsourced Labor*. Minneapolis, and London: University of Minnesota Press.
Whittaker, A. 2009. "Global Technologies and Transnational Reproduction in Thailand." *Asian Studies Review* 33 (3): 319–332.
Whittaker, A., and A. Speier. 2010. "'Cycling Overseas': Care, Commodification, and Stratification in Cross-Border Reproductive Travel." *Medical Anthropology* 29 (4): 363–383.
Yngvesson, B. 2002. "Placing the 'Gift Child' in Transnational Adoption." *Law & Society Review* 36 (2): 227–256.
Zelizer, V. A. 1994. *The Social Meaning of Money. Pin Money, Paychecks, Poor Relief, & Other Currencies*. New York: Basic Books.
Zelizer, V. A. 2011. *Economic Lives: How Culture Shapes the Economy*. Princeton, NJ: Princeton University Press.

Surrogate non-motherhood: Israeli and US surrogates speak about kinship and parenthood

Elly Teman and Zsuzsa Berend

ABSTRACT

Drawing on a comparison of two ethnographic research projects on surrogacy in the United States and Israel, this paper explores surrogates' views about motherhood and parenthood, relationships and relatedness. The paper challenges three myths of surrogacy: that surrogates bond with the babies they carry for intended parents, that it is immoral not to acknowledge the surrogates' maternity, and that surrogacy upsets the moral order of society by dehumanizing and commodifying reproduction. Contrasting the similarities and differences in the voices of surrogates from these studies, the authors argue that surrogates draw on ideas about technology, genetics and intent in order to explain that they do not bond with the child because they are not its mother. This is followed by an exploration of surrogates' definitions of what constitutes parenthood, suggesting that in both contexts, surrogates draw clear boundaries between their own family and that of the intended parents. Finally, it is suggested that surrogates expect a relationship, or a bond, to develop with the intended parents and view their contribution as exceptional moral work which involves nurturing, caring, friendship and solidarity. The paper concludes that for surrogates in the USA and in Israel, maternity, bonding and kin-ties are not automatic outcomes of pregnancy, but a choice. Surrogates in both contexts hold that bonding with other people's children as if they were one's own is wrong while bonding with their couple and creating 'fictive kin' ties with them is the logical outcome of the intense and intimate process of collaborative baby-making.

Introduction

This paper has its origins in the email conversations the authors had for several years, comparing their long-term ethnographic research projects on surrogacy in Israel and the USA (Teman 2010; Berend 2016). The authors found intriguing similarities and differences in how surrogates made sense of the practice and also of the relationship they had with their couple. While the structural conditions and cultural context in which surrogacy arrangements occur are strikingly different in each country, the ways surrogates thought about relatedness, motherhood and relationships had much in common.

In email conversations, the authors also puzzled over the fact that surrogacy is often theorized with so little concern for the actual experiences of surrogates themselves. More specifically, both found it strikingly misguided when critics normatively discuss surrogacy by drawing on cultural 'myths' regarding motherhood and commodification while ignoring surrogates' own understandings of relationships and relatedness. Following Faircloth and Gurtin's (2017) call to examine the intersection of ARTS and parenting cultures through a framework of reflexivity and accountability, the paper explores how surrogates work out meanings and definitions of surrogacy within a reflexive project of family creation through ARTS. Surrogates do this in various ways: in cooperation with the intended parents (IPs), on message boards in which shared ideas about kinship emerge and are crystallized, and in explanations of their involvement in surrogacy to their families, friends and others, within their wider socio-political surroundings and in conversation with local parenting ideologies. The findings indicate that surrogates subscribe to and enthusiastically participate in practices of accountability (Faircloth and Gurtin 2017, 7).

This paper documents practices of reflexivity and accountability by challenging three 'myths of surrogacy.' Myth one, the most prevalent in the scholarship, stems from the naturalistic assumption that mother love and maternal bonding are innately, universally and instinctively rooted in the nurturing relationship between the pregnant woman and her child. Although the culturally constructed nature of bonding theory has been challenged, suggesting that women's emotional attachment to the babies they bear and acceptance of the maternal role are far from instinctive (Scheper-Hughes 1992; Ivry 2010), the assumption that 'normal' women 'naturally' bond with the children they bear is still widespread in surrogacy research (Teman 2010). Scholars assume that surrogates will suffer feelings of regret, loss and depression after relinquishment (Teman 2010). The idea that gestation necessarily creates strong, natural and universal maternal bonds between surrogate and baby also informs many of the policy discussions on surrogacy worldwide (see, for example, Markens 2007).

Myth two is that the surrogate who gestates the baby must be its mother, thus it is not only morally wrong not to acknowledge her maternity, but it could be detrimental to her and the resulting child (Rothman 2000; Anderson 1990). Policy regulations have often followed this logic; current surrogacy legislation in the UK, for instance, considers the surrogate the legal mother of the baby and her husband as the baby's father until a court birth order is awarded. Yet while policy-makers and scholars debate the question of who should be recognized as the mother of the surrogacy-born baby, very few actually ask the surrogates themselves this question.

Myth three is that surrogacy upsets the moral order of society by commodifying and dehumanizing reproduction. Critics claim that the traditional moral framework – in which reproduction is regarded as a 'natural fact' grounded in love, marriage and sexual intercourse (Schneider 1968) – is threatened when bodies, babies, feelings and values are 'for sale' (Ketchum 1992; Anderson 1990). Scholarly and public debate often frame the contractual nature of surrogacy as a sign of the commodification of life and the intrusion of the market into the 'sacred' realm (Berend 2016), adopting the 'hostile worlds' view of market versus family, money versus love and public versus private domains (Zelizer 2000; Berend 2016), pointing out that families have become a matter of choice rather than fate, contract rather than status (Rao 2003). In light of such arguments, policy-makers often view surrogacy as immoral and unethical (Markens 2007).

The paper endeavors to counter these myths and take surrogates' views about parenthood seriously by investigating their kinship understandings. In his *Critique of the Study of Kinship*, Schneider (1984, 200) proposed that scholars take an ethnographic approach and examine 'in detail how the natives themselves conceptualize, define, and describe' kinship and how it is constructed 'in the context of their culture.' Schneider (1984, vii) contended that the existence of kinship 'depends on how it is defined' and that 'its definition does not, nor can it, arise solely as a consequence of "its" real nature.' However, kinship is not simply a matter of definitions but also of actions; it is 'constituted through certain kinds of acts' and is 'carried out in acts that are *meant* and that have meaningful consequences' (Lambek 2013, 246, 247; italics in original).

The argument follows three common findings from the two studies. First, surrogates do not bond with the babies they carry for IPs. They draw on ideas about technology, genetics and intent in order to explain that they are not the mother and thus do not bond with the child. They see gestation as instrumental in 'nurturing parents', the purpose for which they hold themselves accountable, rather than creating kin-ties between themselves and the baby. They do not feel the way many of the scholars who theorize about surrogacy think they *should* feel. Second, while surrogates in the respective studies saw bonding and motherhood in very similar ways, they differ in their definitions of what constitutes parenthood. In both contexts, however, they draw clear boundaries between their own family and that of the IPs'. They do not consider their 'surrobabies' their own children's siblings. Third, surrogates view surrogacy as an endeavor to create babies, parents and families, as a 'journey of love' they ideally share with IPs. In both countries surrogates expect a special bond with their intended mothers (IM). They see their contribution as exceptional moral work which involves nurturing, caring, friendship and solidarity. They are disappointed and hurt when IPs drift away, cut ties, or even worse, redefine the relationship as 'only business.'

By examining surrogates' conceptualizations of relatedness and parenthood in this comparative framework, the paper aims to further the call of Faircloth and Gurtin (2017) to explore the connections between assisted reproductive technologies and parenting culture studies. Several of the central themes of parenting culture studies (see Faircloth, Hoffman, and Layne 2013) – cultural discourses of parenting and family, and especially the centrality of motherhood – shape surrogates' experiences and understandings of surrogacy.

Context and methods

Surrogacy is structured differently in Israel and in the USA. Tightly monitored by the state under a comprehensive law, Israeli surrogacy contracts must be pre-approved by a government-appointed committee. The parties must be Israeli citizens, share the same religion and cannot be related to one another. At the time of the initial study, surrogates had to be single and raising at least one child, while IPs had to be married heterosexual couples, and have no or only one genetic offspring; all surrogates in the 2016 sample were married, as new regulations have permitted married women to become surrogates since 2013. The sperm must be provided by the intended father (IF) and the egg by the IM or an anonymous donor (Teman 2010).

Conversely, in the United States, surrogacy is regulated state-by-state rather than by federal law; state regulation differs greatly. New York, for example, considers surrogacy contracts void and unenforceable. In the most surrogacy-friendly state, California, courts generally reinforce surrogacy contracts; however, there are no statutes regulating surrogacy, and a free market model prevails in which private commercial surrogacy agencies and fertility clinics operate with little oversight. States that allow commercial surrogacy do not limit who can contract a surrogate nor require mandatory contracts or screening, enabling surrogates and IPs to negotiate the conditions of their arrangement through an agency, lawyers or privately. In this state-by-state regulatory situation, agencies, clinics, IPs and surrogates take it on themselves to mandate screening and define and enforce contracts, even when they are not legally binding. Potential surrogates or IPs who are rejected by one agency or clinic may, and often do, find alternatives.

This paper draws from each of the authors' research projects on surrogacy arrangements. While expanded upon elsewhere (Teman 2010; Berend 2016), the following is a brief outline of the methods for each of the studies. The Israeli study included multi-sited ethnographic research conducted between 1998 and 2006, including in-depth, open-format interviews with Jewish-Israeli women who had participated in the process: 26 gestational surrogates and 35 IMs, as well as 20 additional interviews in summer 2016 with surrogates who had given birth in the past two years.

The American study, conducted from 2002 to 2013, employed online ethnographic observation of www.surromomsonline.com (SMO), the largest and most important mediated public surrogacy website in the United States, complemented with email exchanges with 35 surrogates recruited from SMO. Since its inception in 1997, SMO has grown to 30,000 members, both IPs and surrogates, who post stories, questions, information and advice on 30 forums. The overwhelming majority of surrogates participating on SMO live in the USA, and this study contains only US surrogates, thus this sample is referred to throughout the paper both as SMO and as the US sample. Nevertheless, it is important to point out that not all surrogates in the USA participate on SMO. Surrogates on SMO are from many different states across the USA; most are married, are a mix of traditional and gestational surrogates – although more often the latter, and some do both traditional and gestational 'journeys' – who carry for single men, gay and heterosexual couples, frequently from another state or another country.

Both authors adopted the same interpretive and analytic approach and used grounded theory to analyze data. Both authors wanted to understand how surrogates themselves view the meaning of surrogacy, whether these understandings are uncovered through interviews or worked out in online interactions among surrogates. These meanings in turn inform surrogates' actions and have consequences for the practice of surrogacy.

Surrogate non-motherhood

Addressing myth number one, that surrogates bond with the babies they carry for their IPs, both studies found that surrogates adamantly insist to the contrary. Surrogates did not see themselves as the mother of this baby, and most were vocal about never having the emotions that they felt toward their 'own' child, as Israeli surrogate Rinat said: 'It's not mine and that is why I didn't have the kinds of feelings you'd expect.' For some, this lack of bonding was described as 'natural,' just as bonding with one's own baby is viewed as 'natural.' Rinat compared her feelings toward the baby she carried as a surrogate and her own baby: 'You have to understand that the feelings were [in surrogacy] completely different. For instance now (in her subsequent pregnancy) I feel that it is my heart and that it is all mine.' Recalling the first time she saw the baby she bore for her couple, she relayed:

> (I felt) nothing! I also thought that when I saw him I would remember, but nothing... I took him and kissed him and played with him. And it was fun.... It was just like my nephew coming in here, or my neighbor's son.

SMO surrogates also talk about how the 'surrobaby' behaved very differently in utero from the surrogate's own children, and how they nurture and love the baby the way they love a relative's or friend's child. 'It is like I feel toward my brother's child; I love him and protect him but I'm not his mother.' Other SMO surrogates were even more emphatic: 'I have NO feelings of "mother" towards her... I do not in any way feel like I am giving up a child. I am giving a child to her parents after nine months of helping her grow.' In a fairly typical account, a traditional surrogate firmly insisted that she was not the baby's mother:

> This child in my womb is NOT mine. I am nurturing him or her for the time being, but upon birth, he or she will go to their parents where they belong. I am of the belief that the egg used in conceiving this baby was meant for my IPs from before we even knew we'd do this surrogacy. I just don't want anyone thinking this is my baby. I guess I shouldn't care what anyone thinks, as I KNOW what is true.

Both Israeli and US surrogates often compare surrogacy to 'babysitting' – it is a nurturing relationship that entails responsibilities but does not make the child yours. 'I relate my feelings towards this little girl I'm carrying as about the same as I feel towards the little girl I nanny for. I care for her well-being but I have no maternal feelings,' wrote Jenny on SMO. Israeli surrogate Sivan said:

> I would correct people who would ask me, "How will you part from, how will you hand over, how will you give away the baby?"... It was important to me to say that I am not parting from [...]. I am returning something that does not belong to me. [...] It is not mine at any stage. I am a hostess and there was no emotional connection.

Both Israeli and US surrogates firmly believe that the goal of surrogacy is to create families; they see their role as helping the baby's parents to realize their 'dream' and maintain that the child always *belongs* to the IPs. Given that assisted reproduction 'does not automatically assist the making of parents,' rather, it 'assists the making of children' (Strathern 1992, 21), surrogates' emphasis on 'making a family' shifts the emphasis from the baby the surrogate gestates to the family the IPs wish for. The endeavor to 'make families' assumes IPs' relatedness to the not-yet existing children while it also clearly draws the boundary between the IPs' and the surrogate's family.

One way of designating parenthood in both contexts was to 'share' the pregnancy and the birth as much as possible with the IPs, insisting it was 'their' pregnancy rather than the surrogates'. In Israel, where surrogates and couples were able to physically meet throughout the pregnancy because of geographical proximity, surrogates often looked for creative and intimate ways to share the pregnancy as much as possible with their IM (Teman 2010). Israeli surrogate Orna relayed how she tried to help her IM 'feel like she was pregnant' by sharing even the most visceral bodily functions, such as vomiting, dizziness and physical pain with her:

> Look, throughout the pregnancy, I really tried to pass her all sorts of feelings so that she could, so to speak, feel like she was pregnant. I mean, if I had a headache, or I didn't sleep at night, or the fetus was kicking, moving, it pained me or whatever. I would call her and say, 'she kicked me and it hurts.' Then she would say, 'okay, now which side hurts.' I said, 'left. Say it hurts you too.' And she would yell, 'ay, ay, ay' on the phone... and in the beginning when I had dizziness and vomiting, I would call her and say, 'I feel yucky,' and she would say, 'I will go vomit instead of you.' And she really felt like it belonged to her. Like she was pregnant. Because I shared everything with her.

Israeli surrogates also worked out the designation of the pregnancy and of the baby by enabling the IM to participate in all medical appointments and by making sure she 'carried' the responsibility for all the scheduling, filing, and having the necessary documents, and speaking to the doctor. Masha explained:

> 'It was important to me that she be present at all of the ultrasounds. Because it was important to me that she go through the whole experience and that she see the whole experience. [...] I made sure that Tova took part in everything. Because it is really important to me that she go through and feel the whole experience exactly as I do. That is the way I wanted it, that she be my partner, as much as possible.'

US surrogates extensively discussed pregnancy-sharing ideas on SMO; geographic distance often makes frequent physical contact with IPs impossible. When they were physically near the IPs, both Israeli and US surrogates urged their couples to lay a hand on the stomach and feel the baby move, to attend medical appointments and scheduled sonograms to 'visit with the baby.' However, whereas Israeli surrogates were more intent on sharing the manifestations of pregnant embodiment and the experience of pregnancy with the IM, SMO surrogates, even when they lived geographically close, were more focused on sharing their privileged access to the fetus and enabling the IPs to bond with 'their baby.' Some surrogates sent weekly letters to their couple in the baby's name, reporting from the fetus' perspective, similar to the one Liz posted on SMO:

> Hey Mom and Dad it's 'Winston',
> We are officially 16 weeks or 4 months pregnant!!! Only 6 more till I meet you!!! Can't wait. [..] ...Maybe when I see y'all on the 22nd you can hear my heart beating through Liz's tummy!!!!... I've even started growing toenails (can't wait for my big sister to paint them if I am a girl). I am moving around a whole bunch and Liz is starting to feel it especially after she eats!!! She can't feel it on the outside yet but can't wait for you to feel me saying hi to you on the tummy. I am behaving myself and not giving her any trouble.

When IPs lived farther or were unable to attend doctors' appointments, US surrogates often sent ultrasound images and the fetus' tape-recorded heartbeat. Some called their IPs and put their cell phone close to the monitor so that the couple could hear it in real time. Others kept and shared their journals of 'the IPs' pregnancy.' Another popular idea of how to encourage bonding was to play IPs' voice-recorded stories to the fetus so that it would 'know its parents.'

After enabling their couple 'to be as pregnant as possible,' both Israeli and US surrogates wanted to give the IPs privileged access to the birth because it is 'their birth' and 'their baby.' In both samples, surrogates spoke of rituals of participation, and IPs sometimes cut the umbilical cord. In nearly all Israeli cases, the IM attended the birth alongside the surrogate, even if it was a cesarean delivery. Surrogates reported how their IM held their hand, 'helped' push out the baby, and in some instances displayed couvade symptoms in identification with the surrogate's labor. US surrogates also want their couple to be there at the delivery and are devastated when this does not happen. According to SMO consensus, surrogates should include IPs in the birth, and the assumption is that IPs want to be there. Fellow surrogates were incensed when Karen said she did not want her IPs to be present at birth because she was very shy and private. They called her 'selfish' and 'unsympathetic,' depriving her IPs of seeing 'their child born.' 'They may not be carrying the baby but it's still their pregnancy.' The 'sharing' of pregnancy also contributed to, and was reflected in, the intimacy of the surrogacy relationship, as discussed below.

Definitions of parenthood

Addressing myth two, that surrogates are the mother of the surrogate babies, both studies found that surrogates do not consider gestation to be consequential for motherhood, although Israeli and US surrogates have differing views about the role of genetics in relation to maternity. Israeli surrogates firmly distinguished between the potentiality of wombs versus eggs in creating kin-ties and emotional bonds between women and the babies they carry. They considered the egg to be inalienably associated with personhood and self and the vehicle for the creation of kin-ties, whereas the womb was denuded of personal traits and separated from the kin-line. Israeli surrogate Maya, for instance, recalled showing a photo of the twins she birthed to an acquaintance who noted how the babies did not resemble her at all:

> I said "honey, it doesn't transfer through the belly. Genes are genes. It has no connection to me." That is the first question I am asked, if it is my egg. No way, they are not my children. There is no way I would do that. That (using my eggs) would make them my children.

Like Maya, Israeli surrogates were quick to emphasize that the baby they carried for their couple was not 'created from me,' 'is a stranger,' 'is not mine.' They were particularly adamant in stressing that the baby did not share their 'blood' and were nearly all horrified by the idea of traditional surrogacy, which they believed would be 'giving away my own child.' They drew a sharp distinction between gestation and genetics; the former is neutral, while the latter is considered the basis of parenthood.

Israeli surrogates emphasized that they were 'carrying' not just a baby but also the genetic lineage of the IPs. Significantly, even in cases in Teman's study in which a donor egg was used, it was kept a highly protected secret and surrogates, as well as IMs, stressed the genetic link to the IF as encapsulating the continuity of the IPs' kin-line. It was common for the Israeli surrogates to define their role in surrogacy as creating a 'continuing generation' [*dor hemshech*] for their couple, a common cultural idiom that refers not only to a family's continuity, but also abstractly to the continuity of the Jewish-Israeli nation, which is dependent upon its future generations (Hazelton 1977). As Israeli surrogate Orna said, 'This couple has no continuing generation and if they don't have a child then they will simply be erased from this earth.'

Israeli surrogates' insistence on the genetic idiom of a 'continuing generation' must also be viewed in the socio-political context of the Israeli nation's struggle for survival, the reality of war, the biblical imperative, and the collective calling to make up for the murder of six million Jews in the Holocaust (Teman 2010). In this pronatalist context, motherhood is a fundamental status and role for women. Giving birth to a baby for a childless couple is thus endowing the IM with her 'entry ticket' into the collective as well as contributing to the growth of the Jewish population within the context of a 'demographic war' as part of the Israeli-Palestinian conflict.

Such meanings were implied in Israeli surrogates' ideas about their role as surrogates. For example, Neta saw herself as 'doing a moral deed [*mitzvah*]... it is helping a dynasty, the continuation, it is making life continue. That is the biggest mitzvah that can be. And especially since it was a son.' Neta's use of the term 'mitzvah' captures the higher meaning of her contribution. Used in Hebrew to denote the performance of a moral deed and colloquially as an act of human kindness, a mitzvah also refers to the 613 commandments given in the Torah which Jews are obligated to perform as their religious duty. Neta's elevation of surrogacy to the status of a mitzvah connotes the fulfillment of her moral duty as a Jew. Moreover, her indication of the particular social value of a son, who carries on the family name and can also be a combat soldier for the nation, suggests that by 'carrying' the couple's genetic offspring, her surrogacy contributed to the continuity of the collective.

In her discussion of the privileging of genetic kinship in Israeli law and medical practice, Birenbaum-Carmeli (2009, 1023) suggests that genetic relatedness is promoted as fundamental to the Israeli construction of a 'naturalised,' 'gene-based' notion of Jewish identity and the notion of Israel's Jewish collectivity as biologically related kin. She argues that non-genetic forms of kinship, such as adoption, are widely marginalized in Israel by laypersons, medical practitioners, and by the state because they are perceived as a challenge to the 'natural family' paradigm. It is only by demarcating genetic kinship as superior to its alternatives that national interests can be upheld. In view of this metonymic significance of genetics, it appears that for Israeli surrogates, the preservation of genetic continuity is part of an ideological discourse of collective regeneration supported by state political interests; by so vehemently protecting maternity's rootedness in a geneticized notion of 'the natural family,' the Israeli surrogates are protecting the state's interest in preserving a touchstone of the Jewish-Israeli collective.

US surrogates, on the other hand, do not treat the egg as the repository of personhood and kin-ties and do not view genetics as symbols of national continuity. In keeping with US cultural understandings, surrogates regard the family as the basic unit of society and emphasize that good, loving, stable families are the foundation of social progress. In this understanding, planned childbearing is superior to unplanned procreation. Helping people is also considered a positive civic activity. Private charitable volunteering as the solution for social problems has a long history in the USA and not surprisingly, surrogates frame surrogacy as a private charitable act: 'I can't afford to give much to charity right now, and I don't have a lot of free time to volunteer. I figure surrogacy is one way I can contribute positively.' Others often explained that they decided to become surrogates because 'I could feel like I was really making a difference in the world.' These somewhat vague expressions indicate that being useful in the world and helping 'deserving' couples become parents – the public and private sides of 'usefulness' - are connected but not in specific nation-building ways.

Surrogates consider some couples more 'deserving' of children than others; IPs who are successful, stable and do everything to have a baby are worthy of parenthood. 'We found the perfect couple who we know are amazing parents! I have always wanted to help the world become a better place,' Anne explained her decision. Surrogates frequently talk about how unfair it is that young unwed women have babies without even wanting them and how wanted and much-loved babies are precious. In the US socio-political context, this juxtaposition signifies middle-class values and stances about marriage, childbearing, and the merits of planning and hard work; it also implies that young and unwed mothers do not contribute to a better, more stable society.

Surrogates say that 'wanted' children will be happier and healthier adults than 'unwanted' children; thus, surrogacy is the opposite of adoption, as Amy, a traditional surrogate explained:

> To find out that your parents wanted you to be here so much that they were willing to have someone else bring you into this world would, I think, be...easier to hear than your birth parents not wanting you or not being able to take care of you and giving you away.

Genetic parenthood is not a guarantee of good parenthood, surrogates believe, but love, care and preparation for it are: 'Almost anyone can use their DNA to produce a child... not everyone can be a true guardian, friend, or care giver to a child.' Discussions reveal that surrogates see choice as central in both relatedness and relationships; not wanting to be the mother is an ethical choice in surrogacy, the opposite of not wanting one's own child.

Given that genetics and thus eggs are not necessarily central to parenthood and relatedness, and certainly not to nationhood, some surrogates are egg donors themselves, and many others have IPs who use an egg donor. Quite a few traditional surrogates say that they are 'egg donor and gestational surrogate in one,' thus reframing traditional surrogacy in the idiom of donation. Since SMO was the main source of information and support for both traditional and gestational surrogates who discuss and debate issues together, it is not surprising that they have established a common ground regarding surrogacy, one that preferences intentionality.

Since some US gestational surrogates offer their eggs after failed attempts and thus become traditional surrogates and many IPs end up using donated gametes even when they started the process with their own, gestational surrogates on SMO cannot lean too heavily on the fact that they are not genetically related to the fetus because not infrequently neither are the IPs. Indeed, in all of the years of the US study, Berend read only one message by a surrogate who did not want to work with IPs using donor gametes. Yet, whether there is a genetic link to the baby or not, SMO surrogates are adamant that they are 'obviously' not the mother while the IPs are unquestionably the parents:

> Surrogacy revolves around intent. [...] The father of my current IM's future children is a sperm donor. If things don't work out, and she chooses to move on to an egg donor, she will still be the child's mother in my eyes.

IPs' genetic relatedness to the baby is 'supporting evidence' rather than the basis of claims to parenthood.

There is overwhelming agreement on SMO that IPs' desire and love for the baby-to-be is the main unifying feature of surrogacy. In this view, intent, love and parenting make parents, not gestation or DNA. Desire for children coupled with the acts of parenting is a firmer basis of parenthood, according the SMO surrogates, than genetic relatedness.

The surrogate's role, accordingly, is to help couples realize their dream: 'He was created between the love of his mother and father. I was just the middle man.' Thus SMO surrogates, whether gestational or traditional, are in agreement that the babies they carry belong to the couple who desired them and will parent them. Accordingly, 'it shouldn't matter if the child is biologically related to you or not. You went into the agreement with the intention to make a child for another family.' Much like the Belgian parents of Ethiopian adoptees studied by De Graeve and Longman (2013) who believed that 'intensive mothering' of their adopted children makes up for the lack of biological connection, SMO surrogates emphasize 'intended' parenting as a more consequential bond than biology.

By juxtaposing the Israeli surrogates' promotion of genetic kinship with US surrogates' dismissal of genetics in favor of love and care within each country's larger socio-political context of procreation, it becomes clear that definitions of kinship carry cultural and socio-political meanings and that more is at stake in these definitions than meets the eye. It is by carefully defining surrogacy through geneticized ideas about kinship that Israeli surrogates shape their role into one of value to the Jewish collective. For US surrogates, genes can be downplayed because what is important is the IPs' desire to become parents and the surrogates' to help them. Thus, surrogates are fulfilling the IPs' as well as their own private dream to 'make a difference,' creating 'good,' 'chosen' and 'deserving' families.

Faircloth and Gurtin (2017, 8) argue that 'increasingly intensive requirements of... parenting are assumed to be relevant even to the preconception period' and childless couples have to pass intense scrutiny to establish their commitment to parenthood. Our findings show that this scrutiny works out differently in our respective contexts. In Israel, the state takes responsibility for intensively screening IPs to ascertain only 'deserving' citizens – hetero-normative, married couples with no criminal record for whom surrogacy is a 'last resort' to genetic parenthood – can use this option. In the USA, much of this work is done by surrogates, especially when it is an independent arrangement without agency mediation. Beyond the cultural ideas about intensive parenting as proper parenting (Faircloth and Gurtin 2017), the free market model of the social organization of assisted reproduction, with little state oversight and screening, contributes to US surrogates' central role in assessing IPs' 'deserving' status.

In the following, it is argued that for both Israeli and US surrogates, a common definition of kinship emerged that firmly placed the creation of families through surrogacy in the context of intimacy, caring and love. These acts of family creation are morally meaningful for surrogates in both social contexts.

Labor of love

Addressing myth three, that surrogacy necessarily commodifies and dehumanizes reproduction, both studies found that surrogates defined surrogacy as a labor of love with moral consequences and as a relationship of potential closeness and intimacy. In both cases, surrogates seemed to seek out couples with whom they 'clicked' – a concept that both Israeli and US surrogates used to describe the immediate emotional response they thought of as a signal or proof that the relationship 'was meant to be' (Berend 2016; Teman 2010). Surrogates in both settings described their 'journey' in terms that are usually reserved for romantic relationships, applying similar idioms of 'chemistry', 'connection,' and even 'falling in love,' to define the affinity they felt, or believed they felt, upon their first meeting, 'virtual' or in person, with their IPs.

In most Israeli cases, the language of clicking and chemistry was reserved for the IM and excluded IFs. Masha remembered the first time she saw her IM, Tova:

> The minute that our eyes met, there was chemistry. From the minute that we began to speak it felt like I was sitting with someone whom I had known for five years. I knew right then that she was the one.

Chemistry, to the Israeli surrogates, was believed to be the defining element that could make or break a surrogacy arrangement, as Lihi noted:

> Chemistry is something that you can't replace with anything else. It is like having chemistry with your spouse, you either have it or you don't. That's why I say, if I do it (surrogacy) for another couple it won't be the same.

Similarly, SMO is filled with countless excited surrogates' posts, describing the first meeting in the idiom of 'chemistry,' and wishing each other the best by ending their postings with 'May you find your perfect match soon.' Marianne followed SMO advice to 'listen to her heart' and wait for 'the perfect couple,' and was happy to report that she found them: 'It was just like you all said, when you know you just know.' SMO surrogates, who were mostly married women, as well as married Israeli surrogates, spoke about 'clicking' with the couple, although subsequently they usually strove to bond with the IM.

In both countries, references to 'chemistry' situate third-party procreation in the same context as is generally understood for 'normal' procreation, i.e. the result of affinity and love rather than interest (Schneider 1968). Some Israeli and US surrogates spoke of their hesitation whether to 'do it again' for a different couple. As Israeli surrogate Talia explained, 'You only fall in love once for the first time.' Likewise, Mary, who contemplated doing a second surrogacy, posted on SMO: 'I have a very close relationship with [my former IPs], and like a marriage where a spouse has passed, I don't know if I can "recapture" THAT relationship... .'

With few exceptions, both studies found that surrogates often decided between two or more potential couples by relying on intuitive recognition and familiarity. 'Listen to your gut feelings' was an oft-given advice on SMO. Surrogates described their first meetings with their couple as fated, 'love at first sight,' and as a relationship that was 'meant to be.' US surrogates advocate careful research and clear thinking but the 'heart' plays a decisive role in their chosen loyalties to IPs.

Israeli surrogates also described an 'unexplainable connection' and a 'shared fate' that they felt with their IM, and often relayed how they felt 'close to' her or how they 'really love her.' Sally described her thoughts the night before her first meeting with the IPs much like one would describe a blind date:

> I wondered what kind of couple they were...They have to be a couple that get into your heart a little. I am not saying that I have to fall in love with them at first sight, but that they should get into the heart.

In their first fateful encounter, Sally's expectations were fulfilled: 'As soon as they came up the escalator I said "this is the couple"... And that was it. From that moment there was a "click". Just like that.'

Rinat described the 'click' she immediately and intuitively felt with her IM in the idiom of a fated match. She relayed how she had inexplicably recognized her IM Sarit from her distant perch at the window of her second story apartment: 'And they came in the car, and she is looking, and I said to her, "Sarit, I am here." As if I had known her from somewhere… and from then on she decided that I would be her surrogate.' Likewise, SMO surrogate Melissa remembered her first meeting with her couple: 'I fell in love with them the first time we met.' 'I was introduced to only one couple and that is where I stopped, I loved them from the start,' wrote Rene. SMO surrogates often describe the match as one that was destined: 'We must have talked on the phone for hours. Right then I knew this was meant to be.' Surrogacy is considered a 'job of the heart.' Both Israeli and US surrogates often speak about their relationship with their IM by drawing upon various available definitions of social relationships, at times 'like a best friend' or 'like sisters' or even twins. Israeli surrogates also compared their relationship with their IM to that of a 'temporary marriage' and described an intimacy that was sometimes 'closer than a husband and wife.'

Surrogates' conceptualization of their 'journey' as one of shared love is consistent with the understanding of procreation in US – and Western – culture (Schneider 1968). When surrogates evoke their love for the couple, or the couple's love for them, they are highlighting the moral, private, emotional and equalitarian nature of the relationship, elevating it above the contractual arrangement yet distinguishing it from family relatedness. The understanding of surrogacy as a 'labor of love' counters the stigma of mercenary 'baby production' (Berend 2016). When IPs stay aloof, surrogates often excuse their coldness by explaining that infertility makes people guarded; surrogates often hope to heal such 'emotional damage' by being patient and considerate. Surrogates' expectations for real friendship are in stark contrast to descriptions of Indian surrogates who largely have no relationship at all with the IPs (cf. Rudrappa 2015).

Israeli surrogates were fearful of losing the intimacy they created with the IM. Tsila relayed two weeks after she gave birth to the surrogate baby: 'the next day I cried a lot because I thought that I was losing a friend.' Sapir similarly relayed her sense of anxiety over the approaching birth because she thought the IM might abandon her after the baby was born:

> When I gave birth, I became anxious and sort of depressed. Not because of what's-his-name, the baby or something, but because the whole thing was over…. I was really afraid that I was going to lose her…. So I started to cry in the [hospital] room, because it was over. Look, I was the center of everything for a year. And when I was hospitalized, I cried. I would cry at night, until the next day…. I cried over Shoshana, not the baby. So the next day Shoshana came when the welfare officer arrived, and Shoshana said that she didn't want to break it off and that calmed me down…. I can no longer imagine myself without speaking to Shoshana at least once a week. And it calms me down that I have permission from her to call her whenever I want, and even to visit her.

Many SMO surrogates report a sense of loss and sadness after the birth: 'I can feel our relationship changing. The calls are less often and shorter. The distance has begun.' Many surrogates are disappointed, even depressed; others are resigned: 'There is nothing wrong with feeling pain during or after surrogacy. LOVE is painful, no?' Elsewhere (Teman 2010; Berend 2016) we expand upon the anguish surrogates often feel when the relationship with IPs ends abruptly after the birth.

For both the Israeli and the SMO surrogates, remaining in touch after the baby was born constituted acknowledgement that the IPs valued the surrogate's contribution, and that she herself as a person was valued beyond the monetary payment. In both settings, surrogates wanted to believe that these chosen loyalties were 'thicker than blood.' Surrogates in both samples felt immense pride in having been a surrogate, of having given life, and continued to view surrogacy as part of their identity even when IPs were ungrateful.

Israeli surrogate Sally told of her experience two years after her surrogacy when she worked in an office supply store. She told everyone she met about having been a surrogate, and was amazed by the reactions of complete strangers: 'Everyone says 'kol hakavod' [well done], what a mitzvah [moral deed] you have done.' US surrogates were similarly proud of their achievement. Many women said that surrogacy is a good way to teach their children the value of altruism. Surrogates often wrote disparagingly about family members who objected to surrogacy, describing them as 'selfish' and lacking empathy. Some surrogates started 'screw you' threads where women who had some falling-out with family members over surrogacy could post their stories and support one another. Thus surrogates exercised choice in their family relationships even when relatedness was a given. They were the happiest when their IPs affirmed chosen affinities by their continued loyalty: 'they are my real family, even better than my own family.' However, their own and other SMO surrogates' experiences also teach women to uphold the moral value of creating families even in the face of disappointments. Sarah, SMO moderator and surrogate to eight babies, expressed this stance:

> Don't stop giving the love that you so want to give ...Even the most perfect journey is never PERFECT! IPs obviously go into surrogacy with intent of starting a family. Some will keep us in their lives and some will not. ...The beauty of assisting others in creating family is beyond words for me.

Conclusion

Myths surrounding surrogacy guide not only popular perceptions of the practice but also the decisions of policy-makers. Long-held convictions about gestation creating kin-ties and in- utero maternal-infant bonding obscure the empirical experience of the relationships of surrogacy. As Roberts (1998) and Jacobson (2016) have also noted, the time has come to listen to surrogates. The ethnographic data from the two studies suggest that surrogates do not bond with the babies they carry but do bond, or hope to bond, with the IPs. Surrogates do not consider themselves the 'mother' of the baby but want to be appreciated and acknowledged as collaborators by their IPs both during the pregnancy and after the birth. When couples were less than grateful after birth, surrogates often felt betrayed. Both Israeli and US surrogates considered continued contact with the couple as confirmation that the morally significant act of assisting reproduction gives rise to 'ties that bind.'

The studies reveal that for surrogates, whether kinship is understood to be based on genetics (Israel) or on intent (SMO), it is also and most importantly based on love and care. For the surrogates in both studies, maternity is not an automatic outcome of pregnancy; indeed, gestation does not create kin-ties at all. Moreover, bonding with the fetus is revealed as not a universal emotional outcome of pregnancy, but a choice. Surrogates in both contexts hold that bonding with other people's children as if they were one's own is wrong, while bonding with their couple is the logical outcome of the intense and intimate process of collaborative baby-making. 'Fictive kin' ties and a lasting solidarity between surrogates and IPs is the surrogate dream in most cases.

It may be easy to dismiss surrogates' accounts as false consciousness, or point out the rare newspaper article about a surrogate who decided to 'keep the baby,' but if we want to understand surrogacy, we should consider empirical data. The two studies together show that both Israeli and US surrogates consider surrogacy as a morally meaningful undertaking that creates families; surrogates 'nurture parents,' not just 'their babies.' They bond or hope to bond with the IPs rather than the babies whom they do not consider their own. According to surrogates, 'surrobabies' always belong to the IPs; surrogates help create these babies for the IPs' who want to parent them. Ironically, surrogates' interpretations of surrogacy and their actions ultimately uphold the 'sanctity' of the nuclear family rather than undermine the 'moral order' critics see as threatened. We do not argue with the stratified understanding of assisted reproduction (Faircloth and Gurtin 2017, 11) but wish to highlight some genuinely collaborative arrangements within these two national contexts. Surrogates' accounts are important evidence for the ongoing and joint redefinitions and acts of kinship in ways that can be made intelligible in their cultural and socio-political contexts.

Ethical approval

The studies described in this paper received ethics consent IRB approval from the University of California Los Angeles, the Hebrew University of Jerusalem and Ruppin Academic Center.

Acknowledgments

The authors wish to thank the surrogates who shared their stories with them during these research studies as well as Zeynep Gurtin, Charlotte Faircloth and two anonymous reviewers for their comments.

Disclosure statement

No potential conflict of interest was reported by the author.

Funding

The authors derive no financial benefit from any direct applications of their research.

ORCID

Elly Teman http://orcid.org/0000-0001-8353-9965

References

Anderson, E. 1990. "Is Women's Labor a Commodity?" *Philosophy and Public Affairs* 19 (1): 71–92.

Berend, Zsuzsa. 2016. *The Online World of Surrogacy*. London: Berghan Books.

Birenbaum-Carmeli, Daphna. 2009. "The Politics of 'The Natural Family' in Israel : State Policy and Kinship Ideologies." *Social Science & Medicine* 69 (7): 1016–1024.

De Graeve, Katrien, and Chia Longman. 2013. "Intensive Mothering of Ethiopian Adoptive Children in Flanders, Belgium." In *Parenting in Global Perspective: Negotiating Ideologies of Kinship, Self and Politics*, edited by C. Faircloth, D. Hoffman, and L. Layne, 136–150. London: Routledge.

Faircloth, C., and Z. Gürtin. 2017. "Fertile Connections? Thinking across Assisted Reproductive Technologies and Parenting Culture Studies." *Sociology*. doi:10.1177/0038038517696219

Faircloth, C., D. Hoffman, and L. Layne, eds. 2013. "Introduction." In *Parenting in Global Perspective: Negotiating Ideologies of Kinship, Self and Politics*, 1–18. London: Routledge.

Hazelton, Leslie. 1977. *Israeli Women: The Reality behind the Myths*. New York: Simon & Schuster.

Ivry, Tsipy. 2010. *Embodying Culture: Pregnancy in Japan and Israel*. New Brunswick, NJ: Rutgers University Press.

Jacobson, Heather. 2016. *Labor of Love: Gestational Surrogacy and the Work of Making Babies*. New Brunswick, NJ: Rutgers University Press.

Ketchum, S.A. 1992. "Selling Babies and Selling Bodies." In *Feminist Perspectives in Medical Ethics*, edited by H. Bequaert Holmes and L. M. Purdy, 284–294. Bloomington: Indiana University.

Lambek, M. 2013. "Kinship, Modernity, and the Immodern." In *Vital Relations: Modernity and the Persistent Life of Kinship*, edited by Susan McKinnon and Fenella Cannell, 241–260. Santa Fe, NM: School of Advanced Research.

Markens, Susan. 2007. *Surrogate Motherhood and the Politics of Reproduction*. Berkeley: University of California Press.

Rao, Radhika. 2003. "Surrogacy Law in the United States: The Outcome of Ambivalence." In *Surrogate Motherhood: International Perspectives*, edited by R. Cook, S.D. Sclater, and F. Kaganas, 23–35. Oxford: Hart Press.

Roberts, Elizabeth F. S. 1998. "Examining Surrogacy Discourses: Between Feminine Power and Exploitation." In *Small Wars: The Cultural Politics of Childhood*, edited by N. Scheper-Hughes and C. F. Sargent, 93–110. Los Angeles: University of California Press.

Rothman, Barbara Katz. 2000. *Recreating Motherhood*. New Brunswick, NJ: Rutgers University Press.

Rudrappa, Sharmila. 2015. *Discounted Life: The Price of Global Surrogacy in India*. New York, NY: NYU Press.

Scheper-Hughes, Nancy. 1992. *Death without Weeping: The Violence of Everyday Life in Brazil*. Berkeley: University of California Press.

Schneider, D. M. 1968. *American Kinship. A Cultural Account*. Englewood Cliff, NJ: Prentice-Hall.

Schneider, David M. 1984. *A Critique of the Study of Kinship*. Ann Arbor, MI: University of Michigan Press.

Strathern, Marilyn. 1992. *After Nature: English Kinship in the Late Twentieth Century*. Cambridge: Cambridge University Press.

Teman, Elly. 2010. *Birthing a Mother: The Surrogate Body and the Pregnant Self*. Berkeley: University of California Press.

Zelizer, V. 2000. "The Purchase of Intimacy." *Law & Social Inquiry* 25 (3): 817–848.

Ideals, negotiations and gender roles in gay and lesbian co-parenting arrangements

Cathy Herbrand

ABSTRACT
This paper engages with the complex gender and parental dynamics experienced in the context of co-parenting arrangements. These arrangements, based on mutual agreement, involve people who commit to raising a child together, possibly with their respective partners. These family forms are usually pursued to avoid what is perceived as the uncertainty surrounding alternative assisted reproductive options such as donor insemination or surrogacy, and to allow the child to have two biological and sexually differentiated parental figures. This paper explores some of the opportunities and challenges presented by co-parenting by focussing on the experiences and accounts of lesbian women and gay men engaged in such family arrangements. Drawing on work by social theorists of the family, the main characteristics of these arrangements are first examined to show that while co-parenting might first seem marginal, it appears particularly well adapted to contemporary social constraints and parenting expectations. The second part of the paper shows how in practice, dominant gender norms remain largely unquestioned and can lead to tensions and unbalanced parental power relationships between biological parents, which, in turn, can present a significant challenge to these family arrangements. The analysis therefore suggests that while co-parenting offers the potential for parents to reconcile contradictory social expectations and provide them with opportunities to create family practices that suit them, these are restrained by existing gender norms, in particular by the prevailing role of the biological mother.

Introduction

> *From a personal* standpoint, *the sacrifices it would take for me to have a child by myself, using genetic material from a friend or a sperm donor, outweigh my desire to have a child. I can, however, picture an arrangement with a man I love and care about but don't marry, someone who also wants to have a child, organizing from birth the custodial arrangements that divorced and never-married people have been honing for decades to raise children.*
>
> Emily Witt, Future Sex (2016)

Conceiving and raising a child with someone who is not a committed conjugal partner has long been disapproved of and is still regarded by some with concern or disdain, whilst the nuclear family based on love remains a highly valorised model of parenting (Beck-Gernsheim 2002; Gabb and Fink 2018). Despite this context, this way of having children, referred to as 'co-parenting', has increasingly become a family-building option in itself. Originally describing the cooperation of parents in raising their children after divorce, this term now also designates people who deliberately pursue this type of family arrangement and commit to raising a child together, possibly with their respective partners. Co-parenting arrangements can vary considerably, including two to four adults or more, of different genders and sexual orientations (Erera and Segal-Engelchin 2014; Herbrand 2017). While this route to parenthood initially involved mostly gay and lesbian individuals, the rise of specific websites dedicated to such arrangements and the organisation of networking events, such as 'speed-dating' for prospective co-parents (Ravelingien, Provoost, and Pennings 2016), has facilitated its expansion amongst heterosexual women and men in recent years (Jadva et al. 2015). These 'co-parents' – both the intended biological parents and their possible partners – need to determine and negotiate their own roles and responsibilities, as well as the practical arrangements of family organisation.

Despite this expansion of co-parenting arrangements, there is still little research exploring this phenomenon. Current studies have mainly looked at co-parents' motivations and characteristics, showing that such arrangements are usually pursued in order to avoid what is perceived as the uncertainty surrounding alternative assisted reproductive options such as donor insemination or surrogacy, and to allow the child to have two biological and sexually differentiated parental figures (Segal-Engelchin, Erera and Cwikel 2012; Erera and Segal-Engelchin 2014; Jadva et al. 2015; Herbrand 2017). However, there is a pressing need to understand how these new family forms work in practice, in particular regarding parental experiences and gender dynamics. These family arrangements constitute unique sites of parental experimentations where biological parenthood is created and experienced for itself, separately from romantic or sexual commitments. Co-parenting is all the more interesting given that co-parents benefit from significant leeway to negotiate, implement and manage their family arrangements due to lack of pre-established models and medical or legal procedures to follow.

In this paper, I explore some of the opportunities and challenges presented by co-parenting by focussing on the experiences and accounts of lesbian women and gay men engaged in such family arrangements. Drawing on work by social theorists of the family (Giddens 1992; Jamieson 1999; Beck-Gernsheim 2002), I first examine the main characteristics of these arrangements to show that while co-parenting might first seem marginal, it appears particularly well adapted to contemporary social constraints and family expectations, including in relation to intensive parenting (Lee et al. 2014). Co-parents aspire to mix dominant parenting culture with their own reflexive ideas. I then show how in practice, dominant gender norms remain largely unquestioned and can lead to tensions and unbalanced parental power relationships between biological parents, which in turn, can present a significant challenge to these family arrangements. I therefore suggest that while co-parenting offers the potential for parents to reconcile contradictory social expectations and provide them with opportunities to create family practices that suit them, these are restrained by existing gender norms, in particular by the prevailing role of the biological mother.

Methods

This paper is derived from a wider research project on multi-parenthood in Belgium, which focussed on lesbian and gay co-parenting arrangements (Herbrand 2008). Given the difficulty of reaching this hidden group of parents, the study was based on a purposive sample, comprising 26 co-parents with whom I conducted extensive in-depth interviews in 2006 and 2007. There were 11 women and 15 men, aged between 28 and 52 years old and living in Belgium. All but one co-parent were white and the majority were from middle-class backgrounds. I recruited co-parents through snowball sampling, after regularly attending gay and lesbian association meetings and events related to the debate on adoption by same-sex couples taking place at the time in Belgium (Herbrand 2006). The accounts I collected allowed me to analyse in detail nine distinct co-parenting cases, each including two to six co-parents (see table below). Most co-parenting arrangements involved three adults: in two cases, a lesbian couple and a gay man and in four cases, a gay couple and a single lesbian ($n{\sim}2$) or straight ($n{\sim}2$) woman. Two co-parenting cases were also made up of two different child-rearing arrangements and involved three households, i.e. that of a lesbian couple, of their first child's father and of their second child's father. All but one couple's relationships originated prior to the decision to have children. As most of the 11 children included in

Table 1. Composition of the co-parenting cases.

	Biological mother Child 1	Partner of biological mother Child 1	Biological father of Child 1	Partner of biological father Child 1	Biological mother Child 2	Partner of biological mother Child 2	Biological father Child 2	Partner of biological father Child 2
Case 1: 6 co-parents 2 children: Pam and Tom	Anne*	Claire*	Tom*	Steven*	Claire*	Anne*	Patrick*	Sam*
Case 2: 5 co-parents 2 children: Max and Jane	Liz*	Emma*	Daniel*	Anthony*	Emma*	Liz*	Peter*	
Case 3: 4 co-parents 1 child: Maria	Susan*	Charlotte*	Brian*	Mark*				
Case 4: 3 co-parents No child yet	Jessica*	Sophie*	Denis*					
Case 5: 3 co-parents 2 children: Greg and Liam	Paris	Helene	Michael*					
Case 6: 3 co-parents 1 child: Angela	Marta*		John*	Ken*				
Case 7: 3 co-parents 1 child: Alison	Deborah*		Bob	James				
Case 8: 3 co-parents 1 child: Oliver	Zoe*		Jeff*	Richard				
Case 9: 3 co-parents 1 child: George	Laura		Steve*	Fred**				

*Interviewed.
**Not part of the initial parental arrangement.

these cases were very young at the time of the research (i.e. most were under four years old), I did not interview them.

Given the need to capture co-parents' respective views and to understand the negotiations taking place between different members of the family, I asked to meet all co-parents involved in a joint family arrangement, which was possible in most cases, and to interview them separately. I did follow-up interviews with seven of the 26 participants in order to gain a better understanding of the way their experiences of co-parenting had developed over time. All interviews were transcribed verbatim, anonymised and then thematically analysed (Silverman 2001). In order to identify contrasts and overlaps in the accounts, I read and coded all transcripts individually, before analysing them in comparison to the other co-parents involved in the same family configuration (Eisikovits and Koren 2010). This provided data according to thematised individual accounts, as well as allowing a case by case analysis (Taylor and de Vocht 2011). Names, places and identifying details in the interview extracts have been altered.

Challenging parental ideals and expectations

Over recent decades, social theorists have shown how family and intimate relationships have undergone significant changes in terms of definition, forms and functions, and also with regards to how they are idealised (Weeks 1989; Beck and Beck-Gernsheim 1995; Morgan 1996; Smart 2007). In the sphere of intimacy more particularly, Giddens has described how individuals, in late modernity, aim to achieve what he called 'the pure relationship', one based on electivity, auto-regulation, reflexivity and equality. Individuals are perceived to have greater freedom to choose their partner as well as to negotiate the conditions of their relationship, which 'exists solely for whatever rewards that relationship can deliver' (Giddens 1992). According to Giddens, this is part, more broadly, of the ongoing 'democratisation of the private sphere' (1992, 188), which also affects parental, sexual and friend relationships. In line with this theoretical framework, Beck and Beck-Gernsheim have highlighted the central role of love and individual interests in intimate relationships, which simultaneously creates more instability and ruptures as relationships depend on partners' satisfaction.

Jamieson has called, however, for caution and a more complex understanding of these processes, as ideological discourses do not automatically translate in practice. According to her, 'few relationships, even friendships, are mainly simply about mutual appreciation, knowing and understanding' (1999, 482). In particular, she documents how empirical research has shown the persistence of gender inequalities, both in everyday life and at institutional levels. More importantly, Jamieson and other scholars have highlighted how the arrival of children might destabilise the couple relationship since parenting tends to exacerbate gender unbalances (Hays 1998; Jamieson 1999; Sevón 2012; Mannay 2015). Despite discourses promoting equality and shared domestic and parental tasks between both parents, mothers usually remain the child's main caregiver, in particular in contexts where much pressure is currently put on maternal bonding and prolonged breastfeeding (Faircloth 2013).

This 'intensive and embodied motherhood' is part of and contributes to the prevailing middle-class culture of 'intensive parenting' (Lee et al. 2014), which encourages parents to be fully dedicated to the child, reflexive, informed and accountable for their decisions, not only once they become parents but even before conception. Parents are regarded as responsible for the future of their child. As Faircloth and Gürtin point out, 'this rule-bound reproductive and parenting landscape is in many ways contrary to the increasingly fluid, creative and elective landscape of "do-it-yourself" adult intimate relationships' (Faircloth and Gürtin 2017, 6). In addition, 'fathers have not been immune from this trend towards a more "intensive" style of parenting, but it remains mothers to whom these cultural messages are largely targeted, and around women's reproductive choices that the fiercest debates reign' (Faircloth and Gürtin 2017, 4). This can be especially overwhelming and frustrating for mothers, in particular if they value and expect gender equality. It is important to note, of course, that these difficulties in sharing equally domestic and parental tasks are not absent from gay and lesbian families either. While many lesbian and activist works have insisted on gender symmetry and the search for equality within lesbian couples and families (Dunne 2000; Sullivan 2004), Gabb showed how a more nuanced approach is needed in this field and that challenging gender norms does not necessarily involve egalitarian sharing of domestic and parental tasks (Gabb 2005).

Parents, moreover, need to articulate these ideals guiding personal life with those prevailing in public and professional spheres, i.e. performance, flexibility, autonomy and independence. Beck-Gernsheim shows how parents are therefore caught between a number of incompatible demands: being influenced by ideals conveyed by the 'pure relationship', encouraged to be competitive and successful at work, having a satisfying social network and, simultaneously, engaging in intensive parenting (Beck-Gernsheim 2002).

Well-adjusted contemporary families

As illustrated by examples from this study, co-parenting can facilitate the achievement of a number of the ideals and expectations highlighted above, including: the biological heteronormative family model; the possibility to negotiate and implement reflexive parental practices; the need for a committed but personalised parenthood; and the desire for a better and more flexible life-balance.

The primacy of genetic parenthood and the 'mum and dad' family model

As for many individuals for whom conceiving children is not straightforward, either because they are in a same-sex relationship, have no partner or have infertility problems, the co-parents I interviewed had to make reproductive decisions based on several available options. Each of these options had their advantages and drawbacks in terms of practical and legal requirements and implications. Most importantly, each of them aligned or conflicted with particular personal values and family ideals. As Lewin mentions in her studies of lesbian mothers, 'the method a woman chooses for becoming a mother is as vital a part of her objectives as her desire to be a mother. These methods are not only a means to an end, selected for reasons of convenience, but vehicles for crafting statements about the self [...]' (1995, 112).

Lewin's statement can be extended to the gay men and lesbian women I interviewed. It was quite clear that they chose this route to parenthood because of specific family ideals and after giving it much consideration. None of the participants mentioned legal or financial reasons for pursuing co-parenting. In many cases, they could have opted for adoption or assisted medical reproduction, which are permitted in Belgium and could have been more convenient, at least for lesbian couples.[1] In spite of those possibilities, they worked to achieve a family configuration in compliance with their values and identity. As I explain in more detail elsewhere (Herbrand 2017), there were four recurrent reasons given by participants for choosing co-parenting: the desire for biological descent; to experience pregnancy for women, to know the child's background; and most importantly to provide the child with a mother and a father.

Significantly, most co-parents said they would not have been satisfied with adopting a child, as it was crucial for them to have 'a child of their own' with whom they shared biological connectedness. Denis explained: 'I want it to be my child, and by that I mean a biological part of me'. Co-parents' preferences reflected broader family ideals prevailing in a culture valuing and emphasising genetic relationships (Carsten 2004; Nordqvist 2017). This need for a biological child was also closely related, for many women I interviewed, to a desire to share bodily connections with the child through pregnancy and breastfeeding. Most co-parents insisted, however, that they were reluctant to use reproductive medical assistance. Despite valuing conventional parental ideals, this was not a matter of disapproving of the use of reproductive technologies as potentially 'unnatural'. All of them had indeed used self-insemination to conceive their child at home, i.e. some kind of 'assisted' reproduction. This was not an attempt to avoid either medical treatments or clinical settings as such. Rather, most co-parents rejected the use of external genetic material from someone who would not be part of the intended family configuration and play a role in childcare. In particular, female participants in my study expressed different kinds of fears regarding the uncertainties surrounding sperm donor conception. Not only did they want to know the child's medical history, but they also felt uncomfortable imagining themselves or the child possibly meeting the donor or donor-conceived siblings without being aware of it (Herbrand 2017). As Charlotte explained, using an unknown donor would generate too many questions and anxieties that she did not feel ready to face:

> You don't know what is being concocted behind the scenes. You don't know where the sperm comes from, who gives it… there are too many unknown factors, it's too much.

Finding a 'known donor' or an 'accessible surrogate' who could supply the child with information about origins was nonetheless not sufficient for the interviewed co-parents. What many really wanted was for the child to be raised by both biological parents, including the biological father, with whom he or she could develop significant and lasting relationships. Parenthood had to be embodied. This was often part of the family model they might have known themselves and wanted to reproduce. Moreover, as Michael, a biological father, added: 'It makes integration easier. A father and a mother, it is not essential, I think, but if it is possible, it's easier for the child with respect to the outside'.

Although this study focussed on gay and lesbian co-parenting arrangements in Belgium, existing studies (Erera and Segal-Engelchin 2014; Jadva et al. 2015) converge in showing that the main reason individuals, either gay or straight, pursue co-parenting in different contexts is the desire to provide the child with an involved mother and an involved father. Segal-Engelchin et al. (2012) indicate that the single Israeli heterosexual female co-parents they interviewed were also motivated by the practical and financial facilities offered by family arrangements, but their initial motivation was still to maintain 'two-gender parenting'.

The discourses through which co-parents' accounts are constituted reflect a powerful set of ideals associated with biogenetic relatedness, maternal bonding and the heterosexual family. As Faircloth and Gürtin observe (2017), these ideals are still imbued with middle-class values, despite the diversification of family forms. Co-parenting therefore offers the opportunity to achieve these ideals and re-align with the conventional heteronormative family model, while simultaneously challenging parental prevailing norms, such as the single-family household. In this respect, it is important to note that medically assisted options such as use of sperm donation could not entirely fulfil these needs, as they would not provide the child with parental involvement from both biological parents. This is why some individuals, like the gay men and lesbian women in this study, often spent a long time looking for potential co-parent(s) with whom to conceive a child. This reminds us that new family forms do not necessarily require the use of reproductive technologies.

Auto-regulation, negotiation and reflexivity

The co-parents in this study, in particular those who instigated the arrangement, had spent a lot of time considering and planning their family situation. This is not particular to co-parents, as reflexivity and accountability are increasingly expected from would-be parents, in particular from those who wish to adopt or conceive a child through medical assistance. Prospective parents need to demonstrate their capacities to be 'responsible reproducers' and to behave as 'actual mothers and fathers' even before becoming parents (De Graeve 2012; Faircloth and Gürtin 2017). Authors have also shown that gay and lesbian individuals are more inclined to reflect upon and query their parental practices, not only because 'the reproductive process requires couples to make explicit their choices about the family they seek to become' (Nordqvist 2012, 646) but also because they try especially hard to perform 'good parenthood' in environments where gay and lesbian parenting might still be regarded with suspicion (Dunne 2000; Mamo 2007; Donovan and Wilson 2008).

The particularity of co-parenting was, however, characterised by participants as one without intervention from any social or medical institution. Unlike with adoption or medical assistance, co-parents do not have to be scrutinised and follow specific rules or procedures imposed by an external institution. This enabled co-parents to follow their preferences, but also forced them to reflect on what kind of family and parental role they wanted, and how to make these happen in practice. In addition, all aspects of the family arrangement had to be negotiated, agreed on and implemented between both biological parents and their possible partners. These included each parent's responsibilities as well as the practical and ideological orientations of the child's education and upbringing (e.g. the housing arrangement, the financial costs and the choice of school).

Most participants tried, as much as possible, to discuss and plan the important steps and dimensions of the arrangement before the child's conception to avoid disappointment or conflict later on. These decisions could be quite delicate and complex, though, especially when more than two people were involved in the discussion. Claire recalled, for instance, how she, her partner and the child's future biological father talked for a long time about the possibility of baptising their child. Although this was something important to her because of her Catholic background, she eventually gave it up as the other co-parents were opposed to the idea. Similar discussions and negotiations happened, John told me, about vaccination and organic food, and they all had to make concessions to find an agreement. Once the main elements of the arrangement had been agreed upon, it was frequent for co-parents to indicate these decisions in a shared document. This informal 'charter' was used as a means to reflect on and record their choices, rather than as a form of binding contract.

Prior to the birth of the child, co-parents also often invested significant time and effort in getting to know each other better to secure the foundation of the family relationship which they sought to build. For instance, Michael described how he and his lesbian friend couple organised role-playing games between the three of them in order to anticipate and discuss each other's reactions to the child's education and food habits, amongst other things. As Michael said, 'of course, it will be different in real life but it already gave us an idea of how we would react'. They also met other co-parents to understand better how this relationship could work in practice. These mechanisms of anticipation and reflexivity were all the more important since the rise of co-parenting websites had enabled co-parenting between people from different backgrounds and sometimes from different countries (Jadva et al. 2015, 1897). Indeed, while co-parents in four of the nine cases included in this study were friends and knew each other relatively well before discussing the possibility of having a child together, the other participants met through specific meetings or websites for prospective co-parents. One co-parenting was actually taking place across France and Belgium and the co-parents had put a complicated housing system in place to manage the child's circulation between them. In another case, the biological mother recalled how she and her partner were keen to parent with a French gay couple they had met online and found very nice. They nonetheless realised after a few months of contact and interaction that the distance would make the arrangement too difficult to sustain and they reluctantly had to abandon it.

As these examples illustrate, co-parenting cases constituted very reflexive and personalised family arrangements, where individuals were inclined, or at least encouraged, to be open-minded and to demonstrate respect, forward-thinking and conciliation. This reflects more broadly the 'processes of reflexive negotiation' described by Beck and Beck-Gernsheim (2013), whereby individuals need to create and implement their own family practices given the absence of pre-established models and rules for their particular situations. These parental reflexive negotiations are increasingly present and required in a number of families, such as co-parenting ones, whose boundaries and mode of functioning need to be explicitly defined (Hudson 2017).

Personal preferences, adjustable roles and parental commitment

Intention and individual preferences were key in the co-parenting cases in this study. All the co-parents I interviewed insisted on the importance of respecting everyone's personal desire to have children and to consider the extent to which each co-parent wanted to be involved. As mentioned earlier, the possibility of considering and following personal preferences was facilitated by the absence of intervention from any social or medical institution. Parental roles were therefore not left assumed, but were open to reflection, critique and planning.

As for the biological parents, they always wanted the two of them to be actively present in the child's life and to become the legal parents. All biological fathers therefore recognised the child at birth and were highly involved when the child was staying with them, assuming all the usual childcare tasks. While this directly aligned with the family model co-parents valued – one that was based on two sexually differentiated parental figures – and was agreed by all co-parents from the outset, it differed significantly from most donor-conceived lesbian families, where the biological father, if known, usually occupies a more marginal position (Haimes and Weiner 2000; Donovan and Wilson 2008; Ryan-Flood 2009).

The roles and responsibilities of the respective biological parents' partners were subject to more discussion and negotiation, as their desire to have children varied considerably. While most female partners I interviewed shared the biological mother's desire to have children,[2] the male partners' feelings and views about parenting were mixed, some of them stating clearly that they were not ready to take care of a child but would support and accompany the intending biological father in his parental project. Whenever possible, co-parents therefore tried to define parental roles that would fit with the particular needs and desires of each biological parent's partner.

Female partners usually wanted to be very involved with the child, often to the same extent as the biological mothers, either in terms of decision-making or caring activities. Reflecting Dalton and Bielby's findings on lesbian couples (2000), 'the[ir] decision to become parents figure[d] centrally in each partner's commitment to the relationship'. However, unlike what happens frequently in lesbian-headed two-parent families (Dalton and Bielby 2000), female partners in co-parenting arrangements often defined their role as that of a 'parent', insisting that they were not a 'mother'. This role was indeed perceived as unique and distinct because of the biological relationship to the child. Nearly all co-parents I interviewed considered that the biogenetic tie both biological parents shared with the child granted them privileged and specific symbolic parental statuses and thus did not attempt to challenge this by creating equality between both partners. Their initial commitment and their full parental engagement were especially important elements for the partners of the biological mothers and constituted a defining element of their parental identity. None of them wished, for instance, to be regarded either as a step mother or a god mother. Claire, a female partner, explained that adopting the status of a 'god mother' would have facilitated her recognition towards the child, but it would not have reflected her role and her tie to the child in a satisfying way. She also insisted that her role could not be compared to that of a step parent:

> I believe that my role is more important than that of step parents, because, well, I was there since the very start. I am not in a situation that I didn't want. This is a situation that I really wished for. It is very different; I have taken the train at the same time as the others.

The roles and responsibilities of the male partners I met were, however, more diverse and changed over time. They were rarely the ones who instigated the co-parenting arrangement and they were usually less involved with organising co-parenting. For instance, Anthony did not initially wish to be involved in taking care of the child, but he perceived himself as a 'co-father' who took part in the decision and played a secondary parental role in the child's life. As the child grew up, Anthony explained that he became more attached to the child and thus also more involved both in everyday caring tasks and in the long term, as he adapted his will in favour of the child.

Others, like Ken, initially felt a bit 'scared' at first and did not wish to get too involved with the child. He preferred to 'assist' the arrangement, as he said, by supporting the biological father in his role of father and give his opinion about the decisions made between both biological parents:

> I have a 'consultative position' somehow… and this fits me very well. […] I consider myself rather as a 'co-father' than a 'step-father' because there hasn't been any breaking and re-construction. It was John [the biological father]'s desire that I joined. We have shared the adventure together and I have seen the child evolving with his Mum and Dad since the very start.

Again, the early commitment of the biological parents' partners in the family arrangement constitutes a key element that they were keen to emphasise and foreground. It created and marked the specificity of their parental position and connection to the child. If parental commitment was crucial in participants' views, it did not necessarily imply a full engagement in raising a child. In the context of co-parenting, it meant agreeing to be part of the family arrangement from the outset and committing to play a role in the child's life at some level – be it in terms of decision-making, caring activities, advice or responsibilities, as well as to respect and apply the principles and decisions that had been collectively adopted. These commitments were always private and informal, at most recorded in the informal 'charter' mentioned earlier. This document did not have any contractual or legal liability, though.

The importance granted by all co-parents to parental commitment echoes current social expectations for parents to respect and endorse fully the parental engagement and responsibilities they have taken. In Western societies, where contraception and reproductive technologies are often available, having a child is indeed no longer supposed to be – or at least be presented as – an 'inevitable accident', but rather a 'conscious choice' that individuals are required to make for themselves and then take responsibility and be accountable for. In this respect, co-parents affirmed their choice of wanting to have children and being entitled to have them, despite their sexual orientation or relationship status. Moreover, while co-parenting might not initially have represented their ideal option for having a child in other circumstances, it enabled them to create and experience parental roles more in line with their own desires, to adjust these roles to their preferences and needs. This can be seen as an advantage of these family arrangements with respect to more conventional families where respective roles and responsibilities might often be taken for granted and less openly and thoroughly discussed.

Flexibility, work-life balance and child-centred parenting

Interestingly, participants often stressed the unexpected advantages co-parenting brought to their life in terms of time management, because of the possibility of sharing the child's upbringing and residency between the co-parents. This was of course highly influenced by the type of arrangement they would put in place, but again, co-parenting offered a certain flexibility in this respect and unlike divorced parents, co-parents did not have to follow specific legal requirements regarding the child's residency. Besides practical constraints, they could, in theory, opt for any arrangement that suited them. In practice though, most co-parents opted for an arrangement in which the child would live with their mother during the week and would spend every other weekend and perhaps Wednesday afternoons with the father.[3] In one case only, the housing was shared equally between both biological parents. The child's holidays were also usually shared between both parents, and sometimes, some co-parents spending part of their holidays together, i.e. the child with both biological parents and their respective partners.

In two cases where the children were still very young, co-parents took the opportunity to experiment with implementing more flexible and innovative childcare arrangements that would fit with their life and work constraints. One of these cases included both biological parents and the mother's partner. They decided from the outset that they would try to share everything they could equally between the three of them, including expenses, responsibilities and caring activities. In order to achieve this in practice, they met at the beginning of each month and planned with whom the child would be on a daily basis. Although they lived quite far away from each other, they also tried to spend time as the three of them with the child whenever possible. In the other case where flexible arrangements were implemented, the three co-parents had decided to live in the same street and would take turns to go and collect the children at the nursery according to their work obligations.

If this appears complicated, especially when arrangements had to be made between three households,[4] co-parents usually highlighted the advantages of these shared childcare systems, as long as the initial arrangements and principles were respected (see below). Even when they had difficulties in getting used to these arrangements initially, many co-parents, both men and women, said that they eventually greatly appreciated the spare time they had when the child was not with them. It allowed them to better balance the parental, conjugal and personal dimensions of their life, knowing that the child was always in a safe place. Time could be devoted to their children, but weekends or even weeks could be free for themselves. This system also allowed some co-parents who had variable or busy working lives to manage professional and family life better by sharing and adapting their child's housing arrangements to professional constraints. This is why some families decided to repeat the arrangement for the second child. For instance, in one case, the lesbian couple wanted to have a second child, but the first child's father was not ready. The women therefore turned to another gay friend, with whom they then had their second child. They described the second family arrangement as a 'copy-and-paste' of the first one. According to these participants, children complied very well with this. It enabled them to be mobile and to adjust easily to different environments from an early age.

Simultaneously, these arrangements also encouraged co-parents to maximise their time with the child when it was their turn. Both biological parents emphasised how their lives and priorities were then primarily centred around the child. Although they were not subject to any direct supervision or counselling like intending parents opting for adoption or donor conception, I observed how most co-parents, especially the biological parents, tried to fit with norms of 'good parenting', looking for recommendations on the best way to feed children or to provide the most up-to-date care treatments (Lee et al. 2014). They regularly referred to specialised books or forums they had consulted for advice on child upbringing, breast-feeding, maternal bonding etc. Several co-parents also highlighted how co-parenting could offer some educational advantages and contribute to enriching the child's life by providing him or her with different family and cultural environments. For instance, Anne was very happy that the child could learn and speak another language with his father, who was a German native speaker. This appears to be one way of 'optimising' the child's development and opportunities for the future. Achieving 'good parenting' seemed especially important for participants in my study, as they were well aware of the social scrutiny and possible stigma affecting gay and lesbian families. Some of them mentioned, like Michael, how he 'wanted to prove that it is also possible to have a family who works well'.

Re-essentialising gender roles

Co-parenting, which can seem marginal at first glance, is therefore well-adapted to current societal constraints and intense 'parenting culture', while allowing the implementation of the more traditional family model based on genetic parenthood and sexually differentiated parental figures. It presents, moreover, significant advantages for co-parents by offering them some flexibility, a better work-life balance and personalised parental roles. Co-parents' relationships are, however, also shaped by more traditional gender dynamics, which can have significant implications on all aspects of the arrangement.

The primacy of maternal connection

The main reason for unbalanced gendered relationships was that the biological mother's predominant role and presence usually remained unquestioned. As explained elsewhere (Herbrand 2017), the biological mother was indeed perceived as playing an essential role in the child's life, not only because of her perceived 'natural' ability to take care of the child but also because of the special physical bond she had created with the child through pregnancy and breastfeeding, something that the biological father was seen as lacking. According to Steve, a biological father:

> *A child, in the* beginning, *he needs more his mum than his dad, for the feeding, the night... He comes out of his mummy's tummy, it's not to end up on someone else's body, on someone who has no breast.*

Michael, another biological father, also mentioned this 'maternal instinct' when he was explaining how the biological mother, unlike him, had spotted that the child had chicken pox. He added: 'I am not in favour of saying: "this is for the man, this is for the woman", but it is how I have been educated!' Michael's comment reflects a common tension in participants' accounts: most co-parents were simultaneously in favour of gender equality, while essentialising abilities and roles when it comes to parenting.

At times, the importance and priority granted to the biological mother would lead to decisions and situations where men were at a disadvantage, creating possible tensions between co-parents. These tensions were especially perceptible and expressed around the child's housing arrangements. As mentioned earlier, although co-parents had much leeway in this respect, it turned out that the child was usually living with the mother most of the time. While all the interviewed co-parents confirmed they agreed with this arrangement, it appeared, however, that the decisions regarding the housing arrangements tended to have been first discussed and planned by the biological mother or by the lesbian couple, before they talked them over with the child's father. For instance, Deborah, a single mother, explained how when she met Bob, who would become the father of her child, she stated her conditions regarding the child's housing:

> I didn't want to go further in a project where my child could be taken away from me. This really worried me, so I was trying to test the water. [...] Then, when we met, the first thing I said was: 'Listen, you have to know that for me, there is no question about it, there won't be shared custody. It will be [...] every other weekend and Wednesdays afternoons'. As for him, he was being offered a dream on a silver plate.

Things were usually not put forward so bluntly and straightforwardly by the biological mother. Liz, for instance, recalled how she and her partner tackled this issue with the gay couple who were presented with the co-parenting proposition:

> Question: Have they expressed any demands?
>
> Not many actually, not many because we've already had, if you want, a type of well-set model and a well-defined plan, that was half of the holidays and every other weekend. Anyway, we were a little stressed to suggest to them something which we had already worked out in our mind and that we might have had to change. When they agreed, they agreed for everything.

Interestingly, these asymmetrical housing arrangements were generally well received by the fathers, at least initially. Most fathers said they were very happy to have the child even on a part-time basis, since it already offered them, as seen earlier, the opportunity to be present in the child's life and spend some quality time together. Some fathers even preferred that the mother handles the practical aspects of the parenting arrangement. This solution suited Jeff, for instance, who explained that 'even if it isn't the case anymore, I liked going out, among other things. I thought it was fine every other weekend'.

Peter, on the other hand, was more reluctant about this arrangement at first. He feared it would be difficult for him to spend so much time separated from his child. He then spoke to some friends who were already involved in a co-parenting arrangement and asked them how they experienced this situation. They responded that 'it was actually not easy but anyway, it is the only possibility offered to you'. As implied here, as well as in Deborah's account quoted above, co-parenting does not present the same opportunity to men and women who want to have children. It is more difficult for men to turn to adoption or to surrogacy for social, economic and legal reasons, particularly in Belgium, where clinics did not offer surrogacy for single men and gay couples until recently. They are therefore more restricted in their options for having a child. The biological father participants said that when a co-parenting proposition was presented to them, it often constituted a unique and unexpected opportunity to become a father, particularly if they wanted the mother to be involved in the child's life. Consequently, they were more inclined to make concessions over the decisions regarding the family organisation[5] and, like Peter, accepted or convinced themselves that it was a good thing for the child:

> Question: And now, are you still a little apprehensive about the situation?
>
> No, no. On the contrary, I think it is fine because, in retrospect, I realise that it is important for a child to be with his mother in the beginning. Whether it is a boy or a girl, it's important. Let's be honest about that, the link is really strong with the mum from the beginning. And frankly, I believe it is a good arrangement.

Intensive mothering and gender dynamics

The maternal biological and bodily connection was foregrounded not only to justify the chosen asymmetrical housing arrangement but also the adoption of an 'initial transition arrangement' where the baby would stay at the mother's home during the first months and sometimes the first years after birth, before the final agreed housing system would progressively shape up. According to the co-parents, this would allow mothers to breastfeed and to take care of the infant, while maintaining some stability during this period. The father would see, bathe and bottle-feed the child at different times. After a few months, the father would take the child to spend one afternoon at his home, then a whole day, until the child would finally sleep over. The child could therefore slowly adapt to the changes of family environments.

The fathers in this study had mixed feelings about this transition period. Some admitted they had a hard time not seeing more of their child, even it was sometimes a relief not to have to take care of the child during the night. Most fathers eventually tried to accept the situation and focus on its positive sides, as Jeff mentioned earlier. However, in four cases, these tensions increased significantly when the biological mother decided to extend this transition period or asked to modify the initially agreed housing plan.

After giving birth and taking care of the child for a few months, several mothers I interviewed told me they no longer felt ready to separate from the child. They said they had underestimated the length of this period, and that it would be better for the child to stay longer with them. For instance, Susan explained that she had always wanted a father for her future child. She was first very keen to have this child with the gay couple she and her partner Charlotte had met through an online forum. The four of them regularly met before conceiving the child, but once pregnant, Susan started having many doubts about the parental arrangement and their relationship with the gay couple:

> When I became pregnant, I felt that I needed to be just with Charlotte at our place. I suddenly wondered about what I had done. They [the gay couple] were there and I wanted them to disappear. [...] I was thinking: 'I am going to have to give my child, to share her. How am I going to do this?'

Susan no longer felt that she was ready to have the gay couple present in their life – for instance, she no longer wanted both men to attend the child's home birth as she had initially suggested, or implement shared custody as the gay couple expected. In hindsight, Susan believed that she had been too 'euphoric' about the possibility of having a child and regretted not setting out clear deadlines and writing down decisions. She said that she was too afraid that directly addressing sensitive topics could have jeopardised the co-parenting arrangement. She added, while crying, that she was still scared about the way things were going to evolve in the future. While the shared housing arrangement was supposed to start six months after birth, her daughter had still not stayed overnight at the father's home after a year.

By contrast, the single women I interviewed often complained about feeling lonely during their pregnancy. They had expected the child's father to be more thoughtful and present. However, once the child was born, most of them also found it very difficult to separate from the child, and the transition period lasted longer that what had been planned. The need for prolonged breastfeeding was often mentioned as a reason to extend this transition period, but several mothers I interviewed also referred to psychological or paediatric studies on maternal attachment to explain that it was not good to remove the baby too soon from the mother, particularly not overnight.

This situation could cause significant tensions, and all parties suffered. In particular, the fathers who happened to be in that situation told me that they felt prevented from building a strong and steady relationship with the child. This was the case with Brian, the biological father of Susan's daughter mentioned earlier. According to Brian, it was clear from the outset that he wanted to be an involved father and have his child live at his home half of the time.

He therefore felt frustrated not to be able to have her stay over for more than a year. This also made recognition of his new role as a father by family and relatives more difficult, as he was rarely seen with the child at his home or at family events.

These asymmetrical housing arrangements had also tangible consequences for the relationship the partners of the biological parents were able to establish with the child. As mentioned earlier, significant differences were noticeable between the mother's and father's sides in this respect. If the prior existence or absence of desire to have a child clearly influenced one's own parental involvement and commitment, the chosen housing arrangement was of utmost importance too. Indeed, female partners had more opportunities to interact with the child and create meaningful relationships, whereas the restrained time spent with the child did not help the fathers' partners to take a child-rearing role and become more committed to the child, especially during the early months and years.

At the same time, the biological mothers who experienced these problematic situations ended up torn between conflicting ideals: their desire to have the central role in the child's life and their wish to provide the child with an involved and attentive father. While striving to have priority over the child enabled them to benefit from some privileges and to have more control over the family organisation, this made them subject to more responsibilities than the father and his partner. This could be especially frustrating for these mothers, as co-parenting might have been an opportunity to share out parenting tasks and responsibilities, as well as to spare mothers from being the primary care givers. Due to these conflicting ideals, some mothers found themselves caught unwillingly in these pervasive and intensive mother-child relationships, therefore unexpectedly reproducing tensions and difficulties experienced more broadly by many mothers (Thomson et al. 2011; Faircloth 2013; Lee et al. 2014).

However, it is important to point out that these tensions and problems occurred in about half the co-parenting cases involved in the study. Although the maternal and paternal roles tended to be essentialised and the biological mother occupied a privileged situation, it did not necessarily engender difficulties. Many co-parents were getting along well and their arrangement proceeded without major issue, sometimes several years after the child was born and, in one case, even after the separation of the lesbian couple. This might partly be explained by the fact that most co-parent participants had known each other for a long time and had sometimes been friends beforehand. Moreover, even in the problematic cases described above, there was rarely 'open conflict', not only because of the father's goodwill but also because both parents usually wanted to maintain the parental arrangement and to keep collaborating. Their relationships were indeed complex and reciprocal. Most mothers I met wanted a present father in the child's life and, in some cases, wished to have another child with the same father. Maintaining good relationships, however, proved at times more difficult to concretise in practice than these co-parents had expected.

Conclusion

When examined as a parental mode in itself, the characteristics of co-parenting appear to conform particularly well to the prevailing ideals of flexibility, personalised choices, auto-regulation and work-life balance promoted both in contemporary social life and in intimate relationships (Giddens 1992; Beck and Beck-Gernsheim 1995). Co-parents were able to combine dominant parenting culture norms with their own family ideals and parental preferences. While providing the child with the more traditional mother-and-father parental figures, this family configuration also helps to achieve contemporary expectations of intensive, reflexive, committed, child-centred and shared parenting (Hays 1998; Faircloth 2013; Lee et al. 2014). It indeed encourages both biological parents to be actively involved with the child and assume the whole range of parental responsibilities and caring tasks. Of course, as this 'shared parenting' does not happen simultaneously and to the same extent between the father and the mother, it does not offer the 'egalitarian sharing' of parenting tasks that many women want. However, it still provides the mother with the opportunity to alleviate caring and parenting responsibilities, enabling her to keep some time for herself. Co-parenting seems, therefore, to constitute a family route that is particularly well adapted to the 'competing and often incompatible demands' with which individuals are faced in the family and work spheres (Beck-Gernsheim 2002).

Co-parenting did enable a number of participants to reconcile prevailing expectations in the socio-professional domain with those of the private sphere and to find a better balance in their life. This does not mean that they all achieved this in practice or that it works well at all levels. The analysis illustrates how more traditional gender norms continue to prevail and shape co-parents' family dynamics, in particular by essentialising the mother's role and emphasising her responsibilities. It needs to be pointed out, however, that co-parenting as such tends to prompt these gendered processes when it is chosen by individuals who precisely value the conventional two-gender parental references and perceive the maternal connection to the child as specific. Co-parents, whether homosexual or straight, could have opted for another route to parenthood, even if they were single.

If these conventional gender norms did not cause any problem to some participants, they did, however, significantly challenge co-parenting arrangements for others. In these cases, several biological mothers appeared to be somehow 'trapped' by their own ideals, having difficulties combining their need for an engaged father with their desire to develop and maintain an exclusive embodied connection with the child. This shows, again, how gender norms, in particular when it comes to parenting, remain dramatically pervasive, even for people who are seeking and defending egalitarian values in the rest of their life.

Overall, whereas co-parenting arrangements are diverse, sometimes complicated and even conflictual, they might increasingly develop as an attractive family model, in particular for single people, as suggested by Emily Witt's quotation opening the introduction. Co-parenting has the potential to offer significant advantages in modern life and help individuals cope and fit with challenging contemporary parenting expectations. There is undoubtedly a need for more research in this area in order to understand better the parental experiences and dynamics taking place in these family situations.

Notes

1. Most reproductive medical services are publicly funded in Belgium and up to six stimulated cycles are reimbursed until the age of 43.
2. There were two co-parenting cases in which the mother's lesbian partner did not want a child. In one case, the couple ended up splitting up when the child was a year old. In the other, the mother's partner eventually accepted the parental project and became fully involved in it.
3. Belgian school ends at noon on Wednesdays.
4. Two co-parenting cases were composed of respectively five and six adults around two children.
5. In view of the fathers' conciliatory attitude and approval, one might wonder whether gay men with higher demands in terms of parenting involvement would have either given up the co-parenting idea or turned to other parental routes.

Ethical approval

This research was carried out in accordance with the usual protocol around ethical approval (granted by the researcher's supervisor) of the National Belgian Scientific Research Funds (F.R.S.-FNRS), in collaboration with the Université libre of Bruxelles.

Acknowledgements

This study was approved by the National Research Ethics Committee North West, reference 11/NW/0653. At the time of writing Louise Locock was supported by the NIHR Oxford Biomedical Research Centre. The views and opinions expressed therein are those of the authors and do not necessarily reflect those of the HS & DR Programme, NIHR, NHS or the Department of Health.

Disclosure statement

No potential conflict of interest was reported by the authors.

Funding

This research was funded by an ASP research fellowship of the National Belgian Scientific Research Funds (F.R.S.-FNRS), in collaboration with the Université libre of Bruxelles. All research participants have given informed consent to participate in the research.

References

Beck, Ulrich, and Elisabeth Beck-Gernsheim. 1995. *The Normal Chaos of Love*. Cambridge: Polity Press.

Beck, Ulrich, and Elisabeth Beck-Gernsheim. 2013. *Distant Love*. Cambridge: Polity.

Beck-Gernsheim, Elisabeth. 2002. "Reinenting the Family." In *Search of New Lifestyles*. Cambridge: Polity Press.

Carsten, Janet. 2004. *After Kinship*. Cambridge: Cambridge University Press.

Dalton, Susan E., and Denise D. Bielby. 2000. "'That's Our Kind of Constellation': Lesbian Mothers Negotiate Institutionalized Understandings of Gender within the Family." *Gender & Society* 14 (1): 36–61.

De Graeve, Katrien. 2012. '*Making Families*'. *Parenting and Belonging in Transnational Adoption in Flanders*. Unpublished thesis Ghent: Ghent University.

Donovan, Catherine, and Angelia R. Wilson. 2008. "Imagination and Integrity: Decision-Making among Lesbian Couples to Use Medically Provided Donor Insemination." *Culture, Health & Sexuality: An International Journal for Research, Intervention and Care* 10 (7): 649–65.

Dunne, Gillian. 2000. "Opting into Motherhood: Lesbians Blurring the Boundaries and Transforming the Meanings of Parenthood and Kinship." *Gender & Society* 14 (1): 11–35.

Eisikovits, Zvi, and Chaya Koren. 2010. "Approaches to and Outcomes of Dyadic Interview Analysis." *Qualitative Health Research* 20 (12): 1642–55.

Erera, Pauline I., and Dorit Segal-Engelchin. 2014. "Gay Men Choosing to Co-Parent with Heterosexual Women." *Journal of GLBT Family Studies* 10 (5): 449–74.

Faircloth, Charlotte, and Zeynep B. Gürtin. 2017. "Fertile Connections: Thinking across Assisted Reproductive Technologies and Parenting Culture Studies." *Sociology, 0038038517696219.*

Faircloth, Charlotte. 2013. *Militant Lactivism? Attachment Parenting and Intensive Motherhood in the UK and France. 1 edition.* New York: Berghan.

Gabb, Jacqui, and Janet Fink. 2018. *Couple Relationships in the 21st Century - Research, Policy, Practice*. Cham, Switzerland: Palgrave Macmillan.

Gabb, Jacqui. 2005. "Lesbian M/Otherhood Strategies of Familial-Linguistic Management in Lesbian Parent Families." *Sociology* 39 (4): 585–603.

Giddens, Anthony. 1992. *The Transformation of Intimacy. Sexuality, Love, and Eroticism in Modern Societies*. Cambridge, UK: Polity Press.

Haimes, Erica, and Kate Weiner. 2000. "'Everybody's Got a Dad…' Issues for Lesbian Families in the Management of Donor Insemination." *Sociology of Health & Illness* 22 (4): 477–99.
Hays, Sharon. 1998. *The Cultural Contradictions of Motherhood*. New Haven and London: Yale University Press.
Herbrand, Cathy. 2006. "L'adoption Par Les Couples de Même Sexe." *Courrier Hebdomadaire Du CRISP (Centre de Recherche et d'informations Sociopolitiques)* no. 1911–1912: 72p.
Herbrand, Cathy. 2008. "Les Normes Familiales à l'épreuve Du Droit et Des Pratiques: Analyse de La Parenté Sociale et de La Pluriparentalité Homosexuelles." Bruxelles: Université Libre de Bruxelles.
Herbrand, Cathy. 2017. "Co-Parenting Arrangements in Lesbian and Gay Families: When the 'mum and Dad' Ideal Generates Innovative Family Forms." *Families, Relationships and Societies*. https://doi.org/info:doi/10.1332/204674317X14888886530269.
Hudson, Nicky. 2017. "Making 'Assisted World Families'? Parenting Projects and Family Practices in the Context of Globalised Gamete Donation." *Sociological Research Online* 22 (2): 4.
Jadva, V., T. Freeman, E. Tranfield, and S. Golombok. 2015. "'Friendly Allies in Raising a Child': A Survey of Men and Women Seeking Elective Co-Parenting Arrangements via an Online Connection Website." *Human Reproduction* 30 (8): 1896–1906.
Jamieson, Lynn. 1999. "Intimacy Transformed? A Critical Look at the 'Pure Relationship'." *Sociology* 33 (3): 477–94.
Lee, Ellie, Jennie Bristow, Charlotte Faircloth, and Jan Macvarish, eds. 2014. *Parenting Culture Studies. Basingstoke and* New York: Palgrave Macmillan.
Lewin, Ellen. 1995. "On the Outside Looking in: The Politics of Lesbian Motherhood." In *Conceiving the New World Order: The Global Politics of Reproduction*, edited by Faye D. Ginsburg and Rayna Rapp, 103–21. Berkeley: University of California Press.
Mamo, Laura. 2007. *Queering Reproduction: Achieving Pregnancy in the Age of Technoscience*. Durham, NC: Duke University Press.
Mannay, Dawn. 2015. "Achieving Respectable Motherhood? Exploring the Impossibility of Feminist and Egalitarian Ideologies against the Everyday Realities of Lived Welsh Working-Class Femininities." *Women's Studies International Forum* 53 (November): 159–66.
Morgan, David H. 1996. *Family Connections: An Introduction to Family Studies*. Cambridge, UK: Polity Press.
Nordqvist, Petra. 2012. "'I Don't Want Us to Stand out More than We Already Do': Lesbian Couples Negotiating Family Connections in Donor Conception." *Sexualities* 15 (5–6): 644–61.
Nordqvist, Petra. 2017. "Genetic Thinking and Everyday Living: On Family Practices and Family Imaginaries." *The Sociological Review* 65 (4): 865–81.
Ravelingien, An, Veerle Provoost, and Guido Pennings. 2016. " Creating a family through connection websites and events: ethical and social issues ." *Reproductive Biomedicine Online* 33 (4): 522–28.
Ryan-Flood, Róisín. 2009. *Lesbian Motherhood. Gender, Families and Sexual Citizenship*. Basingstoke: Palgrave Macmillan.
Segal-Engelchin, Dorit, Pauline I. Erera, and Julie Cwikel. 2012. "Having It All? Unmarried Women Choosing Hetero-Gay Families." *Affilia* 27 (4): 391–405.
Sevón, Eija. 2012. "'My Life Has Changed, but His Life Hasn't': Making Sense of the Gendering of Parenthood during the Transition to Motherhood." *Feminism & Psychology* 22 (1): 60–80.
Silverman, David. 2001. *Interpreting Qualitative Data*. London: Sage.
Smart, Carol. 2007. *Personal Life. New Directions in Sociological Thinking*. Cambridge, UK: Polity Press.
Sullivan, Maureen. 2004. *The Family of Woman: Lesbian Mothers, Their Children, and the Undoing of Gende*. Berkeley: University of California Press.
Taylor, Bridget, and Hilde de Vocht. 2011. "Interviewing Separately or as Couples? Considerations of Authenticity of Method." *Qualitative Health Research* 21 (11): 1576–87.
Thomson, Rachel, Mary Jane Kehily, Lucy Hadfield, and Sue Sharpe. 2011. *Making Modern Mothers*. Bristol, UK: Portland, OR Policy Press.
Weeks, Jeffrey. 1989. *Sex, Politics & Society. The Regulation of Sexuality since 1800*. 2nd ed. London & New York: Longman.
Witt, Emily. 2016. *Future Sex*. New York: Farrar, Straus and Giroux.

Conceptions of transgender parenthood in fertility care and family planning in Sweden: from reproductive rights to concrete practices

Jenny Gunnarsson Payne and Theo Erbenius

ABSTRACT

It is an oft-repeated trope that the recent medical advances in the field of assisted reproduction have radically transformed the ways in which we can achieve, practice and imagine parenthood. This development has enabled new forms of non-heterosexual family constellations, including same-sex nuclear families and solo-parents by choice, and as a result an increasing number of groups are mobilising politically for access to fertility treatments. Swedish transgender patients are one of these groups; after many years of political mobilisation, they are no longer required by law to go through sterilisation as a compulsory part of gender corrective surgery, and instead today, all transgender patients are offered fertility preservation through gamete freezing. This, in turn, has meant not only that Swedish fertility clinics have faced an entirely new patient group – the transgender fertility patient – but also that the cultural imagination of who can become a parent and what a family might look like is becoming further destabilised. Building on interviews with medical staff, LGBTQ-advocates and complementing qualitative data, this paper seeks to shed light on the very process by which these new rights are translated in the practical context of the fertility clinic, and also what it means, more generally, for cultural imaginations of parenting when a group whose reproductive futures were previously considered either impossible or undesirable are now 'anticipating infertility' and engaging in 'family planning' as central parts of their lifecourse and medical engagements.

Sweden was the first country in the world to regulate what is today commonly referred to as gender reassignment or gender affirmation treatment. *The law on determination of sex in special cases* (*Lagen om fastställande av könstillhörighet i vissa fall (1972:19)*) was instigated in 1972. It required patients to be unmarried, over 18 years old, Swedish citizens, and sterile. In practice, this meant that sterilisation became a normalised part of the treatment process, regardless of whether the patient desired the procedure or not (Author 2 2016). Although we need to consider the historical context at the time – same-sex marriage did not yet exist, and IVF-technology had not yet made fertilisation *ex-vivo* (outside of the body) possible

– it is noteworthy that one explicit reason for sterilisation was to retain a bilineal, gender-binary, and what we today would call *cis-normative* kinship structure. In the preparatory works for the law, the need for sterilisation is explained as being 'necessary to fully eliminate the risk of confusion in the kinship relationships' that would occur if a transsexual person would have children post-surgery (Proposition:6, 50).

Since 2013 sterilisation is no longer a compulsory part of gender affirmation treatment.[1] Instead, this patient group is now offered fertility preservation as part of their treatment, which is subsidised and part of publicly funded healthcare. In combination with more inclusive marriage and family laws, as well as medical advancements in the field of assisted reproduction and fertility preservation, these legislative changes have significantly improved the possibilities for transgender people to become genetic parents. Building on interviews with medical staff and complementing qualitative data, this paper seeks to shed light on the very process by which these new rights are translated in the practical context of the fertility clinic, and also what it means, more generally, for cultural imaginations of parenting when 'anticipating infertility' and 'family planning' become central parts of the treatment for a group whose reproductive futures were previously considered either impossible or undesirable.

Aim and objectives

This paper investigates how, from the perspective of medical professionals and transgender advocates, the abolished sterilisation requirement and changing cultural imaginations about transgender people as (actual or potential) parents have led to new practices in the clinical setting of reproductive care. Specifically, the paper explores on the one hand the process by which the medical staff came to change their conceptions of transgender reproduction and parenting, and on the other how the professionals perceive the remaining obstacles for full implementation of transgender reproductive rights, including fertility treatment, family planning and parenting. By drawing on interviews with professionals working with issues of transgender reproduction and parenting and contextualising qualitative data, it seeks to complement existing research and official reports on patients' perspectives (e.g. Bremer 2011; von Doussa, Power and Riggs 2015; SOU 2017:92; Armuand et al. 2017). Thereby, the paper contributes novel insights towards how transgender fertility care and family planning can be further improved, and move closer to the goal of fully implementing these newly won reproductive rights in concrete clinical practices.

Methodology

The empirical data consists of seven interviews with seven professionals[2], as well as secondary data, including an observation of a half-day of lectures on transgender issues, analysis of two governmental reports, a website on transgender reproduction and parenting, information materials and medical forms. Interviewees were recruited by what best can be described as 'strategic snowballing'. Initial contacts were made with two 'key informants', one at the unit for transgender care and one at the unit for reproductive medicine at the Karolinska University hospital in Stockholm. The key informants provided useful information about other professionals engaged in transgender care, sometimes including specific contact details. All interviewees were contacted by e-mail, and received a description of the research plan before agreeing to take part in the interview. The interviews lasted between 30 and 80 minutes (on average around 60 minutes). The interviewees have been purposefully selected to construct a dataset that provides a broad range of professional perspectives on transgender reproductive care in contemporary Sweden (Gray 2002: 101). All interviews were recorded, transcribed and coded in MAXQDA.

Interviewees, ethical considerations and anonymisation

All seven interviewees are central professionals in the field of transgender rights, fertility and/or parenthood. All of the medical professionals work at the Karolinska University Hospital in Stockholm at one of the two units involved in treatments related to transgender reproduction (the unit for transgender care and the unit for reproductive medicine). The two other professionals work with family matters at the Swedish Federation for Lesbian, Gay, Bisexual and Transgender Rights (RFSL), a national non-profit organisation working for LGBTQ-rights.[3]

Cecilia Dhejne, who is consultant and specialist in psychiatry and sexual medicine at ANOVA (the unit for transgender care) has been interviewed twice: once in 2015 by the first author and again by the second author in 2017. The other interviewees, 'Orvar' (consultant in andrology and sexual medicine), Kenny Rodriguez-Wallberg (consultant and specialist in obstetrics and gynecology at the unit for reproductive medicine), 'Beatriz' (midwife at the unit for reproductive medicine), 'Agneta' (associate professor and specialist in gynecology at the unit for reproductive medicine), and Veronica Berg Hulthén and Anna Nordqvist (RFSL) were all interviewed by the second author in 2017. All interviewees were interviewed individually, except for Veronica and Anna, who were interviewed together. Cecilia, Kenny, Veronica and Anna have been active in public debates on transgender reproduction and are therefore, unlike 'Orvar', 'Beatriz' and 'Agneta', not anonymised. All interviewees have consented to be cited for publication, and have been given the opportunity to read a previous draft of this paper to approve their quotes. The study was partly financed by the Swedish Research Council. The study was conducted in accordance with the principles of the CODEX website (Vetenskapsrådet) and Södertörn University.

Background: Transgender reproduction and non-cis parenting

A review of existing research including 51 studies from Europe and North-America (none of which were conducted in the Swedish context) shows that it is common for transgender individuals to be parents, but proportionally less so than in the general population. It also shows that transgender parents are more often subjected to discrimination in contact with authorities (e.g. adoption processes, issues concerning custody) (Stotzer, Herman and Hasenbush 2014; also cited in: SOU 2017:92, 244). There is also a growing number of studies on transgender fertility preservation, mainly from medical and bioethical perspectives, arguing for increased access to fertility medicine (see e.g. Murphy 2012; Nixon 2014; Chen et al. 2017). One study discusses barriers to high quality healthcare for pregnant men in the US-context, recommending improvements such as electronic medical records and staff training, and identifies the needs for increased knowledge about the post-partum period (e.g. the need to tackle feelings of isolation) (Berger et al. 2015). Another US survey-based study on transmen's experiences of pregnancy and birth confirms that this patient group often desires to have children and are prepared to achieve this goal by going through pregnancy, but also that they express a wish for better information about options for fertility preservation and treatment, as well as access to trans-friendly healthcare providers (Light et al. 2014).

A number of studies analyse the cultural representations of pregnant men, and in particularly of Thomas Beatie[4] – a transman who has gained international recognition precisely for his pregnancy – have been published. Damien Riggs has analysed expressions of 'mundane transphobia' in the Oprah Winfrey talk show, whereby Winfrey in interviewing Beatie utilises a variety of gender-normative rhetorical tropes that renders Beatie as 'not quite' a father and 'not quite' a man (Riggs 2014). Jasper Verlinden has argued that although the significance of this episode of the Oprah Winfrey show is sometimes overestimated, it was indeed a 'watershed moment' for cultural representations of male pregnancies and transgender identity (2012: 109). As we shall see, Beatie's international visibility also played a crucial role for the medical professionals interviewed for our study. Riggs has also analysed transmen's self-representations in online media such as blogs, concluding that their narratives demonstrate that it is possible to renegotiate cultural conceptions of pregnancy and fatherhood and that biology and identity are not necessarily connected. Therefore, he argues that health care staff 'must acknowledge how gender norms shape individual health care needs', and take into account both similarities and differences between pregnant women and pregnant men (Riggs 2013). Michael Toze has discussed how the 'cultural unthinkability' of pregnant men has impacted UK healthcare policy, for example by encouraging hysterectomy, even though sterilisation has never been explicitly required by UK legislation (2018).

There is still a lack of qualitative interview studies on transgender reproductive care and parenthood. One notable Canadian example, however, is offered by James-Abra et al. (2015), who interviewed 40 transgender individuals and their partners. In this study, only two couples described their contact with medical staff as predominantly positive, insofar as that the latter had offered a warm and trans-friendly clinical environment and used gender-neutral terminology. The negative experiences reported by the majority of respondents included the use of cis- or heteronormative language and assumptions, problems with documentation, and being denied certain treatments. Thus, in its conclusions, this study called for staff training, and recommended that transgender patients be included in the process of improving care. In the European context, UK Sociologist Sally Hines (2006, 2007) has conducted important qualitative research on transgender individuals' partnering and parenting practices, the latter related to coming out processes in relation to already existing children. Henry von Doussa, Jennifer Power and Damien Riggs (2015) have conducted an interview study with Australian transgender adults, some of whom were parents. They found on the one hand that normative understandings about gender and parenthood shaped the imaginations and desires of parenthood of their interviewees, but that on the other these norms were re-negotiated and contested by them.

There is one study to date that has investigated the issue of fertility preservation (freezing of eggs) for transmen in the Swedish context.[5] The study reports that the 15 respondents did experience some distress in relation to fertility preservation: because of the associations made between reproductive organs and tissue and the gender assigned by birth, the procedures involved negatively impacted some patients' gender dysphoria. The fact that fertility preservation delays or temporarily stops other treatments (due to waiting lists and/or pausing testosterone treatment) often added further distress. The study found, however, that the patients had coping strategies to help them to go through the treatment. It also recommended staff to use gender-neutral vocabulary, so as do decrease the risk of negatively impacting the patients' gender dysphoria (Armuand et al. 2017).

Theoretical starting points: Biomedicalisation, de-naturalisation and re-conceptualisation

Each new technology of assisted reproduction constitutes a 'dislocatory break' with existing hegemonic discourses of reproduction and parenthood (Laclau 1990; Tinnerholm Ljungberg 2015, 225). Although some of these 'breaks' are more 'dislocatory' than others, what they have in common is that they in one way or another break with culturally dominant ideas of what reproduction 'is' and what forms of reproduction are possible. Therefore, they also have the potential to de-stabilise which forms of reproduction are considered 'normal' and 'desirable'.

As a consequence, in order to become established as commonly used practice, every new invention is necessarily followed by processes of what Charis Thompson calls *normalisation* and *routinisation* of medical practices (Thompson 2005, 80). In Sweden, and many other places, assisted reproductive technologies that were once considered 'unnatural' or 'undesirable' are now legally regulated and routinised in fertility clinics and maternity wards, including non-heterosexual forms of reproduction such as donor conception for lesbian couples and single women. Following this, the biomedicalisation of reproduction has destabilised 'the dichotomy between heterosexual and homosexual experience and the institution of the family, the grounds upon which dichotomous gender is reinforced and maintained' (2007, 3). But, as implied in the term normalisation, each such process generates a new 'normal' as well as new norms (Thomson 2005, 80).

One such new norm that has been discussed by scholars of biomedicalisation is the increased emphasis on health (rather than illness) and individual (rather than public) responsibility that has followed in its wake (Clarke 2003, 171–2). With regards to fertility preservation specifically, this tendency has been acknowledged in relation to so-called 'social egg-freezing' whereby women freeze their eggs in anticipation of age-related infertility, a phenomenon Lauren Jade Martin (2010) has called 'anticipated infertility'. By extending fertility treatment to groups of people that are not actually yet diagnosed as infertile, she argues precisely that there has been a discursive shift in emphasis from 'reproductive illness' to 'reproductive health'. With these developments, she argues, it 'also becomes an *obligation* of those experiencing anticipated infertility to ensure their future reproductive capacity' (2010, 531, emphasis added). Similar criticisms of fertility preservation as contributing to a 'reproductive imperative' have been raised by Jack Halberstam, who has questioned the queer radical potential of Beatie's story, arguing that it rests too much on a 'universalist' narrative that normalises 'everyone's' desire to become a parent (2010, 78). Indeed, as Faircloth and Zeynep (2017) argue, the current culture in both assisted reproductive technologies and parenting is one that requires increased reflexivity and accountability from users, who must not only engage with increasing medical and technological expertise but also negotiate their relationship to normative expectations. For the purposes of this paper, and considering the long struggle to abolish the requirement for sterilisation as a 'natural consequence' of gender affirmation treatment, we argue that the risk of such a reproductive imperative is less relevant to our study than the need for increased recognition of transgender individuals as (potential) parents – and hence (potential) fertility patients.

From normalisation to routinisation in transgender reproductive care

As Author 1 (2013) has argued in a previous study, a crucial political strategy in the debates that ultimately led to abolishing the requirement for sterilisation was to normalise the desire of transgender individuals to have their 'own' children, most notably by drawing on hegemonic and culturally recognisable discourses of parenthood (especially motherhood) and involuntary childlessness, as well as the publicly acknowledged condemnation of previously practiced forms of involuntary sterilisation. As we shall see, the same two themes are prevalent in Cecilia's memories from the time when she herself started to question the legitimacy of the sterilisation requirement and instead began to reconceptualise transgender patients as potential parents.

'A normal family': Normalising transgender reproduction

Cecilia, who has worked with transgender patients since 1999 and today is an outspoken advocate for transgender reproductive rights, describes that the issue of reproduction was mainly considered a 'non-issue' for her and her colleagues in the clinic until the years around 2005–2006. Prior to this time, it was simply so normalised that gender affirmation treatment would automatically exclude future reproduction.

However, as organisations such as RFSL began to raise the issue of sterilisation politically and a new governmental investigation was awaited, things also started to change in the clinic. One notable difference from previous years, she says, was that around 2005, an increasing number of patients seeking full treatment (including surgery) began procrastinating on sending in their applications for changing their legal gender to the Legal Council. They wanted to 'wait a little'. 'And' explains Cecilia, 'finally we began to understand that they were waiting for a new legislation'.

A specific event that she recalls, and comes back to in both of our interviews, is a memory that perhaps can best be described as a 'watershed moment'. This is a meeting in the clinic, whereby a parent of one of her patients asked whether the planned surgery would mean that she would lose the prospects of ever becoming a grandmother. (Meeting patients' parents can be part of the diagnostic process.) When parents started to ask 'Will I not be able to become a grandparent?', she realised that she had to respond 'No, you can't, because we have a *law* that states that your child is not allowed to reproduce.' 'That's what I was supposed to say! That's what I *had* to say!', she emphasises.

One thing that stands out in her account is the affective grip of the generational narrative, whereby sterilisation becomes de-normalised and re-described as an undesirable or even unacceptable practice that breaks generational lineage and introduces an unnecessary loss in the family (the loss of future grandchildren). But this process of de-normalising sterilisation, she says, was not sudden, but emerged gradually over time. She describes how she and the rest of the staff working at the unit for transgender care in Stockholm began to discuss the consequences of abolishing sterilisation in their regular meetings. The possible emergence of pregnant men was discussed specifically. Initially, many had thought the idea to be strange, and questions had cohered around the possibility of a transman 'still being trans' if accepting or desiring to be pregnant. Would a person wanting to be pregnant still 'really' be a man?

Interestingly, another 'watershed moment' described by Cecilia involved a visit by Thomas Beatie, who coincidently was invited to give the opening speech at Stockholm Pride in 2011.

Some members of the medical team heard the speech, and when a patient suggested that they should try to arrange a meeting with him, many medical professionals showed an interest. In the end, around 15 medical professionals gathered to meet with Thomas Beatie in the RFSL building. This meeting is described as having made a big difference for Cecilia and her colleagues insofar as it helped those in the team who had been sceptical to understand that in fact, the issue of male pregnancy was not necessarily so strange. Significantly, this meeting served a crucial role in normalising male pregnancy. Meeting Beatie's family made a big impact on the team. Cecilia recalls that 'they had three children, and those kids were running around and seemed just like any other little rascals (*busungar*)', again demonstrating the normalising effect of describing families of transgender persons as being 'just like any other family'.

Cecilia also remembers that Beatie had 'explained very patiently to' them that they had 'wanted children and his wife could not [carry] them because she had had her uterus removed' and that 'he might not have loved being pregnant but it was important for them to have genetically related children, so he decided to put up with it'. The description of Beatie's choice to be pregnant as a pragmatic one made at least in part because his wife could not be pregnant, further adds to the sense of 'normality', as it alludes to dominant cultural assumptions that the desire to have children is more or less universal and that, in the wake of the biomedicalisation of reproduction, there is a range of ways to fulfil this desire.

Cecilia also emphasises the importance of personal meetings more generally as crucial for the process of normalising transgender fertility treatment. She acknowledges the difficulty for a cis-person to empathise with a transperson's feelings in relation to their body and explains that it may be different for those professionals who work with transgender patients in their everyday work life: 'we learn to understand this through our patients', she says. She continues to explain that 'this thing about pregnancy was extremely difficult to envision [to many]; it is not out of spite [that they are/were skeptical], but it makes an enormous difference if one has met [transgender] people [or not], for the personal sense of it.' She also describes this year, 2011, as not only the year when the issue of sterilisation in gender affirmation care began to be seriously debated in public, but also the time that she and some of her colleagues began working more actively for a law change, in parallel and sometimes in collaboration with RFSL who offered their expertise on LGBTQ reproduction.

New routines: Integrating fertility treatment into transgender care

In the clinical context, much of the initial work of 'translating' the 'new normal' of transgender reproduction to concrete routines and habits cohered around dealing with the organisational and policy structures of the hospital. The dilemma the clinicians faced before the law change was that technologies for fertility preservation were available in the fertility unit, but the legal restrictions meant that they could not formally refer their patients there – although they could inform them of its existence and services. In a sense, we can say that this phase consisted of a form of tactical and informal routinisation. As a result, some patients managed to freeze their gametes before the new law was instigated. 'Beatriz' (midwife at the fertility clinic) reveals that there was one man who just before she was employed there managed to conceive using his own frozen eggs. 'And that was before the law change actually', she adds. When asked further about the period before the law change, 'Beatriz' explains that transwomen had frozen their gametes in the clinic before: 'Yes, I know that, they have had access to the sperm lab for a long time.'

But it was only after the sterilisation requirement was actually abolished in 2013 that the work of constructing a new care chain to fully integrate fertility preservation into transgender affirmation treatment could begin – and this was by no means a straightforward process. It involved, among other things, negotiations with the fertility clinic including knowledge exchange between the different units concerning the medical and psychosocial needs of this specific patient group. 'Orvar' describes how at this time, transgender fertility patients were still an unknown patient group in Swedish fertility care: 'for them, this was like turning a whole lot of conceptions upside down and that was very important, to give them the chance to understand this patient group'. He describes the importance of education for staff, because 'if one has not had a chance think through and be confronted with [new, non-normative] ways of thinking, they can be experienced as provocative, or very odd'. Much of this work was conducted in cooperation with RFSL, members of which participated in meetings with staff and leadership of the fertility clinic to inform them about the specific situation and needs of transgender patients, and assisted with practical advice (including how to design forms that were neither hetero- nor cis-normative). As the work progressed, an increasing number of people became involved, among them representatives from RFSL who according to Cecilia 'did a massive job!'. The patients, she says 'did a real pioneering work by explaining [to other medical professionals], without being frustrated, [...] And now I think it's running very smoothly'.

In addition to the purely medical aspects, this work involved a large number of psychosocial and cultural issues of everyday hospital life and organisation. Examples of this include staff learning to avoid expressions that could be perceived as what Riggs (2013) refers to as 'mundane transphobia' by learning new habits of using the right pronouns, changing the colour and design of forms to be filled in by patients, and posters and images of gametes and reproductive organs. Cecilia remembers for example that 'it turned out that the fertility units' papers [forms] were extremely heteronormative and binary, [...] that they were pink [for egg-freezing]', but also that the staff at the fertility unit had no to little experience of transgender patients and their needs.

Important needs to consider in the new routines included the effect that fertility preservation might have on the patient's gender dysphoria. Both Kenny and 'Beatriz' explain that the very procedures involved in fertility preservation may cause distress because they involve reproductive organs and cells that are culturally – and for many emotionally – associated with the gender with which a transperson dis-identifies. Kenny explains that the use of vaginal ultrasound rod therefore can be emotionally difficult for a transman. Therefore, an important part of the treatment process is to explain why using this instrument is necessary to 'get as close as possible to the ovaries through the vagina'. It is also important that clinicians understand that it can take a long time (even months) for some patients to become emotionally prepared for this procedure, and it is equally important that patients are aware that this is not an obligatory part of gender affirmation care. 'Beatriz' explains that fertility preservation may also cause distress for transwomen, who might not be comfortable with the procedure of depositing sperm. Some of the female patients decide against it for this reason.

Lagging routinisation

One pressing problem for this patient group is the long waiting time involved for both gender affirmation care and fertility preservation. In fact, the long waiting times constitute a crucial reason for some patients to decide against fertility preservation. Anna, who works with issues LGBTQ-families at RFSL empasises that 'this is really something that needs to be tackled' as 'it can be a *long, long, long* wait in every step of gender affirmation care.' She expresses a worry that 'if we continue down this line, in five, ten years, we will have testimonials from people who had wanted [fertility preservation], but that because of the waiting lines have not received it.' Although, as her colleague Veronica says, waiting times is a problem in fertility care more generally, Veronica emphasises the 'added layer' of time and waiting lists for transgender patients specifically, because fertility preservation in itself adds another step to the process, and thereby delays gender affirmation treatments that the patient might feel are equally or more urgent.

'Orvar' speaks of the long waiting times as a result of a 'lagging phase' in healthcare, caused by the proportionally vast increase of patients entering the system in a relatively short period of time. Cecilia also confirms that ANOVA (the transgender care unit) have had an increase of remitted patients of about 30-50% each year: 'In the year of 1999, 20 persons were remitted [in the Stockholm region] for medical investigations and treatment of gender dysphoria, a number that last year [2016] had increased to as many as 350' (Dhejne 2017, see also Dhejne et al. 2014). Among young people, the increase after the sterilisation requirement was abolished is even starker, especially after 2013. Remittances to the unit that diagnoses young people under 18 years of age, have increased from only single cases each year to 197 individuals in 2016, with a dramatic increase since 2013 (Frisén, Söder and Rydelius 2017, 1).

'Orvar' explains this general increase as probably caused by a combination of factors, including increased visibility of transgender persons and issues in mainstream media, and increased possibilities of finding information and communities online, and most likely better treatments and the abolished sterilisation requirement. He also thinks that one possible explanation of the vast increase in the Stockholm clinic specifically might in part be a result of the high quality and good reputation of the clinic, and the fact that it attracts patients from other parts of the country. According to 'Orvar', unfortunately, the increase of patients has not been followed by a proportionate increase in financial resources. This, in turn, has led to what 'Orvar' calls a 'lagging phase' in transgender care:

> It has gone very quickly, the increase. And that healthcare has, what shall I call it, a 'lagging phase' before it is possible to swallow new large patient groups. And such an increase is hard to keep up with. It also depends on the fact that our examination and treatment has been pretty attractive [among patients], I think, so that people from other parts of the country also come here. [...] ('Orvar')

In order to implement a sustainable working pace, he says, the clinic has had to develop a certain acceptance of waiting times in order to ensure a reasonable working environment for the staff.

> It's difficult to keep up, and you can't pressure the staff too much either. You can do a little more, but you can't do enough and then it's just as well to see that 'this level we can do, people do not get burnt out and manage to do a good job'. And then unfortunately it is the case that there are waiting lines, but it's still better than if we would have worn out everyone here within two years. ('Orvar')

The lack of adequate increase of funds is felt at the unit for reproductive medicine as well. Both 'Agneta' and Kenny points out that it is a problem that they receive 'new patients but not more funding'. Kenny emphasises that 'It's a fact that we don't have any more doctors, but we have new patients that need resources, so well... we have done as well as we can'.

Crucially, waiting times for transgender patients are further exacerbated not only as a result of an increase in patients in the clinics, but also because the very extra step of fertility preservation takes time in itself. In addition, fertility preservation has consequences for when hormonal treatments can begin – hormonal treatments that are often urgently needed for psychosocial reasons. Cecilia emphasises that '[Waiting lists] are long! I don't know. Sometimes [my colleague] says that it's two-three months, but it feels like half a year, at least', and explains that 'And then they can't begin hormonal treatment, so there are a lot of people who decides against [fertility preservation]. Although it is possible to begin hormonal treatment and temporarily stop taking then for fertility preservation, Cecilia points out that there are some uncertainties about how this might impact fertility, but adds that 'But we don't know how well they do in the freezer either, to be honest.'

Based on the narratives of the medical staff, the necessity that the legal recognition of transgender reproductive rights needs to be followed by adequate increase in resources. And that the long waiting times required to access this care are not only onerous for patients, but also deter some of them from seeking fertility preservation, which they may have desired under better practical circumstances.

From fertility preservation to a reproductive life-cycle perspective

Significantly, the obstacles for transgender individuals to become parents are not confined to available treatments, the quality of care or even psychosocial difficulties. Indeed, current regulations on assisted reproduction as well as family law are both strictly organised around a cis-normative understanding of kinship and family. As a consequence, transgender people who wish to use their gametes in the future might not be able to do so.

For example, a transman who is not partnered with a person who can gestate a child (e.g. a transwoman, or a cis-woman without a functioning uterus) will not be able to reproduce with the help of a surrogate, since surrogacy-related treatments are not legally permitted within the Swedish healthcare system, and gametes cannot be exported to another country. When asked what a transman could do in this situation, gynaecologist 'Agneta', who works in the fertility unit, simply responds that 'Then you would have to store your sperm abroad.' On the question of whether this is part of her information to the patients, she answer that she does if they ask. In the past, she says, they conveyed more information, but nowadays, 'most of the time they are already informed, they come here as quite young'. She explains that the patients are already well informed when they come for fertility preservation: 'when they come here now, they have gone through their examinations, and then they have the right to this, so it is nothing that we question.' Their role, she says, is more like 'an egg-harvesting and sperm-saving unit', but adds that

> Well, we can… sometimes it's possible to… ask if they have a partner, and of course if you have a transman here that has a [transwoman as partner], then, at least I normally inform them that today, according to the law, you have no possibility to use. Or, I mean with a male partner. That today there is no possibility to use [the sperm]. ('Agneta')

The situation described above is only one of many examples of obstacles that this patient group might face when wanting to become genetic parents. And for those who do manage to reproduce, family law continues to complicate the matter.

Lagging legislation

While 'Agneta's' impression is that the patients are well informed, Veronica, who worked with educational issues in RFSL around the years of 2012–2013, discovered that there was an acute lack of information about different ways for non-heterosexual and transgendered people to become parents.

In her quest to address this lack, she received funding for the project *HBTQ-familjer* that lasted between 2014 and 2017. Her colleague Anna, who also participated in the interview, joined the project in its final year and is at the time of the interview back at her job at RFSL national office, where 40% of her time is spent working with family matters. Veronica and Anna talk about how the family project expanded over time, and included a broad range of activities, including meeting places for LGBTQ families and their children, preparatory family courses, theme-days and -evenings, activities during Stockholm Pride, and the development of a website. They also educated professionals, such as midwives and medical doctors, participated in the media and took part in various referral groups. In short, the website expanded to offer resources and information for people outside of the hetero- and or cis-norms to navigate a system that is based on a cis-normative understanding of parental kinship.

> It's like you say, the basic analysis is that regulations of fertility care and law, [and] family law, is adjusted after [the] cis-, hetero-, nuclear family, so to speak. Which means that the situation for LGBTQ-people become rather special, and we want to gather all information in one and the same place, [...] practically, legally, and medically, as it turned out. And it has to be fact checked and updated. [...] It's for that target group, LGBTQ-persons, who are or want to become parents, with an emphasis on want to, really...

The rationale behind the website is to gather all information about not only possibilities and rights, but also obstacles, possible solutions and instances that may not yet have a solution. While RFSL does work politically on removing obstacles for this group, Veronica argues that the website seek to respond to the concrete questions: 'Which possibilities are there? How can you do it? Which rights do you have, and which ones are lacking?' She states that 'It can be a nuisance that some rights are lacking, but you need to know!'

Legal regulations are central to both the website and the internal discussions held within the project. Veronica says that many people they speak to feel that they are 'tricking the legal system', either because they end up in a situation that they either planned for, 'or [a situation] that only occurs because the system looks like it does, and, yes, [these situations] can be very peculiar'. Anna continues:

> Yes, exactly! We talk a lot about legal security, and the legal security of children. And that there can't be so much security if it is not possible to predict what will happen to the child. If the system is so complicated – which [Swedish] family law is – so that one... Yes, one will be able to make a choice, or to know at least, so that one can make some kind of informed decision.

For transgender people who have changed their legal sex, the legal consequences for the child are often unpredictable. One issue is whether the child will have one or more legal parents from birth, another how parental status is confirmed.

On the website, a special section is dedicated to the current family law and its consequences for transgender parents. It initially states that 'In the current situation, it is not obvious how family law shall be interpreted when it comes to determine parental status for persons who have changed their legal sex', and explains that family law has not been updated after the law on sterilisation was abolished. It points out that although the right to parental designation is not yet mentioned in family law, a court has set a precedence that a person who has changed their legal sex does have the right to receive a parental status that is compatible with their legal sex. That is, a person who is legally recognised as a man has the right to be registered as the father, and a person who is legally recognised as a woman has the right to be legally registered as the mother (regardless of the gametes contributed to the conception of the child). It also confirms that a person who gives birth to a child is always considered its legal parent. Swedish law applies the principle of *mater semper certa est* – which applies equally for any pregnant person. Because of this, the issues of legal parental status for trans persons mainly concern people who are or wish to become parents but have not been or will not be pregnant.

Based on a range of different scenarios that vary depending on the gendered couple constellations (e.g. same- or different sex relationships, contributions of gametes or gestation), the website pedagogically guides the reader through the complex legal consequences in each case: (i) 'Legal woman married to or co-habiting with a legal man'; (ii) 'Legal woman married to or co-habiting with a legal woman'; (iii) 'Legal man married to or cohabiting with legal man'; (iv) 'Person who contributes with sperm', and; (v) 'Person who neither carried the child nor contributed with sperm'. Based on the number and severity of legal complications that may arise in the different constellations, it is quite clear that although both family and marriage law are no longer explicitly heteronormative, they remain fundamentally cis-normative in the way that gestation and eggs are seen as intrinsically tied to women, while the assumption is that only men can contribute sperm.

There is also a lack of knowledge about regulations in the healthcare system. Anna states that:

> There are probably quite a few people who have an idea about the regulations, but don't necessarily know where to find [the information], black on white. There are several people from this group that have turned to us and said that 'I know that it is supposed to work like this, but where can I find [the information] black on white so that I can put it under somebody's nose?' And that has been really difficult to find. And really, [the information that] exists is [to be found] in the [materials] on gender affirmation care or [materials] on gender dysphoria. And often one looks for it in the guidelines for assisted conception, where it is not mentioned at all, even though there are newly formulated guidelines about parts of the assisted conception aspect, of it, where the rules are not at all that explicit. So that's very problematic.

Anna and Veronica point out that even when recent legislations in the field of assisted reproduction were created, these were still formulated in a cis-normative manner. When single people gained access to treatment with donated sperm in 2015, for example, it was explicitly formulated for single *women*, meaning that e.g. a transman seeking treatment with donated sperm would not fit into the available description.

These problems are also acknowledged by the report *Transpeople in Sweden* (2017), which like Anna and Veronica argue that the main problem of current legislation (also recent ones) on assisted reproduction and parental status is that it presupposes a connection between sex-cells and legal gender. Retroactively, it can be argued that previous law changes that have been made to include same-sex parents and single women constitute a set of lost opportunities to adjust the law in accordance with the range of possible parental kinship positions that have been made possible by existing biomedical advances in fertility medicine. For instance, while existing legislation does include the gender neutral position of 'parent', this has only been included as a possibility for lesbian couples; the fact that a child can now have a 'mother' and a 'parent' – but not a 'father' and a 'parent' will not least cause problems when a person changes their legal sex after having had a child. This, in turn, has major consequences for these parents to exercise parental duties that require a legally recognised parental status.

From anticipating infertility to reconceptualising parenthood: Concluding reflections

As stated earlier, the biomedicalisation of reproduction has made possible a whole new set of possible ways for humans to reproduce – one of which is the possibility to preserve fertility for the future. This development has been co-constitutive of new claims to reproductive rights, also among groups who were previously seen as 'inherently infertile'. In the last few decades, transgender patients have been reconceptualised from being seen precisely as non-reproducing subjects to becoming a new patient group with reproductive rights. As this paper shows, however, reproductive rights and access to fertility treatments are worth far less that they ought to be unless they are followed by provisions not just to become a parent, but also to have their specific psychosocial and medical needs met and to practice parenthood in a legally safe way.

As our study demonstrates, the process of reconceptualising parenthood so as to include transgender people has not been a straightforward process. Indeed, the initial work to de-naturalise sterilisation and work for a new law was subsequently followed by a long process of re-naturalising transgender patients as (potential) fertility patients and (potential) future parents. While the interviewed medical staff all expressed positive attitudes about welcoming transgender patients in the fertility clinic, it is clear that they had to find strategies of un-learning cis-normative ways of talking and thinking about reproduction and parenthood as well as creating new routines and habits to implement this new perspective. Crucially, the medical staff and transgender advocates acknowledged a number of remaining obstacles to implementing these new reproductive rights in practice. Such identified obstacles include inadequate funding to meet an increased number of patients as well as cis-normative legislations that either prevent patients to use their frozen gametes and/or create an unsafe legal situation for their children.

Based on these insights, we call for a renewed political debate about financial resources for transgender healthcare (including fertility treatments) as well as a thorough reworking of legislation of assisted reproduction and family law. In line with the governmental investigation (SOU 2017:92), this paper argues that many of the problems can be avoided if legislation is formulated around 'reproductive function' rather than legal sex in order to better mirror existing biomedical advances. This would not only decrease discrimination and difficulties of transgender individuals who are also parents, but it would also increase the legal security of children born as a result of transgender reproductive rights. Moreover, sex education and family planning practices need to be revised along the same lines to become more inclusive: as an increasing number of young people identify as trans- and non-binary, it is crucial also for this group to receive adequate and realistic information about the possibilities for and obstacles to forming a family in the future – and for the cis-gendered majority to gain a better understanding of the multiple modalities of gendered parental positions, so as to prevent discrimination of transgender parents and their children.

Notes

1. In public debates the now obsolete law is often recognised as belonging to Sweden's dark history of coerced sterilisations (for a more detailed account, see Author 2013). A proposal for a new legislation concerning monetary compensation to transgender individuals who have been sterilised was handed over to The Council on Legislation (Lagrådet) on 9 November 2017 (Lagrådsremiss 2017).
2. One interviewee has been interviewed twice, and one interview was conducted with two interviewees together.
3. The organisation was formed in 1950, and gathers 7000 members. It is a national organisation with 38 local branches across the country.
4. Thomas Beatie, a transman from Oregon, USA, first received international media attention for his pregnancy in April 2008. Although Beatie is sometimes referred to as the 'first pregnant man in the world', Currah notes that even at this time it was 'not unheard of in the transgender community' (Currah 2008).
5. The Swedish situation for people undergoing gender affirmation treatment has also been investigated thoroughly by Signe Bremer, who analysed personal narratives of going through gender affirmation treatments before the aforementioned law change (2011). Studies on the experiences of Swedish transgender fertility care and parenting experiences are, still few and far between, which at least in part can be explained by the fact that the law changed only recently.

Ethical approval

The study was conducted in line with the principles of the CODEX website (Vetenskapsrådet) and the authors conformed to the full ethical principles of qualitative research. All interviewees have been given the opportunity to read and approve the accuracy of their quotes before final publication, as agreed at the time of the interview.

Acknowledgements

The authors would like to thanks the editors, the interviewees and the three anonymous reviewers for useful feedback.

Disclosure statement

No potential conflict of interest was reported by the authors.

Funding

The first author's work on this paper was generously supported by Vetenskapsrådet (The Swedish Research Council), DR 344-2012-5209.

References

Armuand, G., C. Dhejne, J. I. Olofsson, and K. A. Rodriguez-Wallberg. 2017. "Transgender Men's Experiences of Fertility Preservation: A Qualitative Study." *Human Reprodution* 32 (2): 383–390.

Bremer, Signe. 2011. *Kroppslinjer. Kön, transsexualism och kropp I berättelser om könskorrigering.* Gothenburg: Makadam förlag.

Chen, Diane, Lisa Simons, Emelie K. Johnson, and Barbara A. Lockart. 2017. "Transgender Preservation for Adolescents." *Journal of Adolescent Health* 61 (1): 120–123.

Clarke, Adele, Janet K. Shim, Laura Mamo, Jennifer Ruth Fosket, and Jennifer R. Fishman. 2003. "Biomedicalization: Technoscientific Transformations of Health, Illness and U.S." *Biomedicine, American Sociological Review* 68: 161–194.

Currah, Paisley. 2008. "Expecting Bodies: The Pregnant Man and Transgender Exclusion from the Non-Discrimination Act." *Women's Studies Quarterly* 3–4 Fall/Winter 2008: 330–336.

Dhejne, Cecilia, Katarina Öhberg, Stefan Arver, and Landén Mikael. 2014. "An Analysis of All Applications for Sex Reassignment Surgery in Sweden, 1960-2010: Prevalence, Incidence, and Regrets." *Archives of sexual behaviour* 43 (8): 1535–1545.

Dhejne, Cecilia. 2017. *On Gender Dysphoria.* Doctoral dissertation, Stockholm: Karolinska Institute. Available: https://openarchive.ki.se/xmlui/bitstream/handle/10616/45580/Thesis_Cecilia_Dhejne.pdf?sequence =1&isAllowed=y (last accessed 24 May 2018).

Faircloth, Charlotte, and Gurtin Zeynep. 2017. "Fertile Connections: Thinking Across Assisted Reproductive Technologies and Parenting Cultural Studies." *Sociology*, online first, 1–18.

Frisén, Louise, Söder, Olle, and Rydelius, Per-Anders 2017. Kraftig ökning av könsdysfori bland unga. Tidigt insatt behandling ger betydligt bättre prognos. *Läkartidningen*, February 2018.

Gray, Ann. 2002. *Research Practice for Cultural Studies: Ethnographic Methods and Lived Cultures.* London: Sage Publishing.

Laclau, Ernesto 1990. *New Reflections on the Revolution of Our Time.* London & New York: Routledge.

Hines, Sally. 2006. "Intimate Transitions: Transgender Practices of Partnering and Parenting." *Sociology* 2 (40): 353–371.

Lag (1972:119) om fastställande av könstillhörighet i vissa fall. (Law (1972:119) on determination of sex in special cases) Stockholm: Socialdepartementet.

Lag (2013:405) om ändring i lagen (1972:119) om fastställande av könstillhörighet i vissa fall. (Law on amendment of the law (1972:119) on determination in special cases. Stockholm: Socialdepartementet.

Light, Alexis, Juno Obedin-Maliver, Jae M. Sevelius, and Jennifer L. Kerns. 2014. "Transgender Men Who Experienced Pregnancy After Female-To-Male Gender Transitioning." *Obstetrics & Gynecology* 124: 1120.

Martin, Lauren Jade. 2010. "Anticipating Infertility: Egg-Freezing, Genetic Preservation, and Risk." *Gender & Society*, 4 (24): 526–545.

Murphy, Timothy F. 2012. "The ethics of fertility preservation in transgender body modifications." *Journal of Bioethical Inquiry*, 9 (3): 311–316.

Proposition. (1972:6). *Kungl. Maj:ts proposition med förslag till lag om fastställande av könstillhörighet i vissa fall. (Proposition (1972:6) HRH's proposition with proposal of a law for determination of sex in special cases.* Stockholm: Regeringskansliet.

Stotzer, Rebecka L., Judy L. Herman, and Amira Hasenbush. 2014. *Transgender Parenting: A review of Existing Research.* Los Angeles: The Williams Institute.

Riggs, Damien 2014. "What makes a man? Thomas Beattie, embodiment, and 'mundane transphobia.'" *Feminism and Psychology* 24 (2): 157–171.

Tinnerholm Ljungberg, Helena. 2015. *Omöjliga familjen. Ideologi och fantasi i svensk reproduktionspolitik.* Stockholm: Stockholm University.

Transpersoner i Sverige: Förslag för stärkt ställning och bättre levnadsvillkor 2017. (Transpersons in *Sweden*: Proposals for strengthened position and improved conditions) SOU: 2017. (Governmental investigation).

von Doussa, Henry, Jennifer Power, and Damien Riggs. 2015. "Imagining Parenthood: The Possibilities and Experiences among Transgender People." *Culture, Health & Society* 17 (9): 1119–1131.

Commentary

Marcia C. Inhorn

In this unique special issue of *Anthropology and Medicine*, guest editors Zeynep Gürtin and Charlotte Faircloth bring to fruition an intellectual conversation begun three years ago at the 2015 American Anthropological Association annual meeting. There, for the first time, they united anthropologists working in two disparate fields—parenting studies and the anthropology of reproduction. The papers delivered in that cross-cutting session demonstrated two important, but taken-for-granted facts: first, that people utilizing assisted reproductive technologies (ARTs) do so in the quest for parenthood; and second, that ART studies usually end at the point of conception, thereby leaving us to ponder how post-ART parenthood unfolds.

As Gürtin and Faircloth have argued since then in several path-breaking publications (Faircloth 2013; Faircloth and Gürtin 2017a, 2017b; Gürtin 2014, 2016), scholarship must attend to the ART-parenthood nexus, for people's pathways to assisted reproduction always involve imagining, accounting for, and attempting to achieve hoped-for motherhood and fatherhood. Desires for parenthood persist around the globe, with nearly 95% of the world's adults telling survey researchers that they want to have children (Inhorn and Patrizio 2015). Children are needed for social and economic reasons. But they are also desired for the joy, meaning, and purpose they bring to parents' lives. Understanding this 'child desire' (Inhorn 1996, 2015) is critical. It underlies the 'fertility-infertility dialectic' (Inhorn 1994), or why infertile people often resort to ARTs (if given the chance), especially in high-fertility societies where children and parenthood are socially expected (Inhorn 1994, 2015). Indeed, in the 40 years since the birth of the world's first test-tube baby, Louise Brown, in England, ARTs have spread around the globe, with infertile couples becoming the parents of nearly 8 million IVF babies (De Geyter 2018).

That ARTs 'make parents' is the key insight forwarded by feminist technoscience scholar Charis Thompson (2005) in her seminal volume, *Making Parents: The Ontological Choreography of Reproductive Technologies*. In that book, Thompson shows that men and women willingly engage in complex 'ontological choreographies' in their often herculean efforts to become parents of IVF offspring. Yet, as Thompson (2002) pointed out in another classic essay entitled 'Fertile Ground: Feminists Theorize Infertility', much of the feminist scholarship on ARTs ignores these parenting desires. Instead, most feminist scholars have focused on the potential harms (social, physical and economic) brought on by new ART interventions. Although this feminist critique of ART risk is clearly justified, Thompson argues that we must also attempt to understand why these technologies are so important to people—what potential benefits and rewards the technologies offer. Ultimately, Thompson advocates for committed feminist ethnography that makes sense of the human desires, motivations, investments, struggles, joys and subjectivities of ART users themselves, both men and women.

To that end, this special issue is exemplary, and builds on Gürtin and Faircloth's work to bring the scholarship on assisted reproduction and parenting into dialogue with each other, including an earlier journal special section entitled 'Making Parents' (Faircloth and Gürtin, 2017b) after Thompson's insight. In this special issue of *Anthropology and Medicine*, each paper is based on intensive ethnography or in-depth interviews, mostly undertaken with less studied populations of ART users. These include: (1) single women who use sperm donation to become 'solo' mothers by choice (the paper by Susanna Graham); (2) infertile women who use donor eggs to conceive with their male partners (the paper by Kathleen Hammond); (3) straight and gay individuals and couples who embark upon costly overseas surrogacy arrangements (the paper by Ingvill Stuvoy); (4) surrogate women who explain why they undertake this form of reproductive labour (the paper by Elly Teman and Zsuzsa Berend); (5) lesbian women and gay men who choose to conceive and co-parent together (the paper by Cathy Herbrand); and (6) health professionals who provide fertility preservation services (namely, sperm and egg freezing) for the increasing population of transgender clients (the paper by Jenny Gunnarsson Payne and Theo Erbenius).

Ultimately, these very complex reproductive arrangements—involving egg and sperm donation, commercial gestational surrogacy, gay and lesbian parenting, and transgender reproduction—are designed to enable heretofore unprecedented pathways to parenthood. Yet, in almost all of these cases, empirical evidence of how these reproductive arrangements are sustained and experienced remains remarkably thin. Foregrounding and filling in these empirical lacunae is very important, especially in cases where we know too little about potential reproductive risks, as well as successful outcomes (e.g. ART use among transmen and women).

Through careful and sensitive ethnography, the papers in this special issue attempt to provide the empirical scaffolding to account for *why* people make particular reproductive and parenting decisions and *what* the consequences might be. As Gürtin and Faircloth argue in the Introduction, such reflexivity and accountability are critical, given that conception is increasingly ART-mediated and parenting decisions are ever more complex. Moreover, the parenting of children created through ARTs has often been described as particularly 'intense'. This is partly because ART children often arrive in 'multiples' (i.e. twins, triplets and beyond), but also because ART offspring are deemed especially 'precious' to their parents after years of infertility, treatment, and sometimes heartbreaking agony.

As shown in the emerging ethnographic literature (Faircloth 2013), parenting in general—and not just of ART children—is becoming ever more involved, time-consuming, expensive, committed, competitive, and demanding, creating for parents heightened pressures, external judgments, and feelings of guilt. In this era of 'intensive parenting' (Faircloth and Gürtin 2017a), it is important to understand that reproductive decisions—including about whether or not to bear a certain kind of ART child—may be rife with anxiety and worries. Calling this 'anxious reproduction', Faircloth and Gürtin (2017a, p. 13) note:

> Paradoxically, while both ARTs and parenting expertise aim to 'assist' reproductive agents in their conception and child-rearing endeavors, they also generate new choices, burdens, responsibilities and accountabilities. The result, unsurprisingly, is increasing anxiety for

> parents and intending parents. In this moment of 'anxious reproduction', despite diligence and the use of multiple resources, reproductive agents often end up feeling overwhelmed, scrutinized and 'not good enough' parents…alert[ing] us to the unrealistic, and counterproductive, expectations currently placed on reproducing the next generation.

Several of the papers in this issue reflect upon these anxieties, with study interlocutors worrying about their own decisions and the ramifications of such decisions for future offspring. For example, Graham's study of 'single mothers by choice' highlights the self-reflexive questioning that many women undertake when contemplating solo motherhood via sperm donation. Concerns about being 'selfish' for wanting a biological child, and the risks of raising a child without a father, are women's key ethical dilemmas. Similarly, in Hammond's paper on egg donation, we see how women struggle to reconcile their own infertility when 'genetically linked' motherhood is typically valorised. In many cases, women experience grief and worry over the perceived 'conjugal inequality' of their own absent genetic links, ultimately trying to distance themselves from their use of donor eggs. As seen in both studies, such decisions to use donor gametes are not easy in a world where biogenetic relatedness is still often upheld as paramount, despite increasing reproductive options.

In addition to heightened anxiety, this turn to ART-assisted conception entails tremendous parental investments—physical, financial and emotional. This may be especially true in cases of commercial gestational surrogacy. Stuvoy's paper in particular focuses on monetary investments—how the costs associated with surrogacy add up, with clinics offering price lists, cost breakdowns, formal contracts, and schedules of reimbursements. Sometimes emptying their bank accounts—for example, Stuvoy's study participants spent between US$35,000 and $400,000 on their overseas surrogacy arrangements—the intended parents of surrogate children rarely view these outlays as 'buying a baby'. Rather, they see these expenditures as the beginning of ongoing investments in their own parenthood and their future children's lives. Although ART baby-making has been thoroughly criticized for its high profit margins and commercialization—as shown most clearly in Debra Spar's (2006) influential volume, *The Baby Business: How Money, Science, and Politics Drive the Commerce of Conception*—Stuvoy's account foregrounds the readiness, self-sacrifice, and careful calculations undertaken by intended parents, both men and women, gay and straight, together and alone.

Surrogates, too, make substantial investments, even though their own self-reflexive accounts are rarely foregrounded in the surrogacy literature. In the interesting cross-cultural study by Teman and Berend, the authors use their ethnographic research with surrogates in the US and Israel to produce a key finding. Namely, surrogates do not see themselves as 'bonding' with a 'baby of their own' (which they must then painfully 'give away'). Rather, they seem themselves as 'making parents' of other couples, with whom they hope to forge and maintain lasting friendships. For surrogates, producing other people's families is a morally meaningful act, often done as a labour of love and care for a couple who cannot otherwise have children. In short, Teman and Berend highlight surrogates' emotional investments, moving beyond feminist critiques that tend to frame surrogacy experiences in terms of exploitation.

A question raised by this and other papers is whether ARTs are somehow transgressive in making 'new' kinds of parents and families. The contribution that most directly addresses this question is by Herbrand, whose ethnographic study focuses on gay men and lesbian women who conceive and co-parent together. On the one hand, these co-parenting arrangements can entail as many as four co-parents and two or more residences, thus constituting new, non-heteronormative, non-nuclear family structures. On the other hand, as Herbrand's study shows, nearly all the co-parents she interviewed considered the biogenetic link between parent and child, and especially mother and child, to be sacrosanct, thereby reinscribing traditional kinship ties and privileging maternity above paternity in parenting.

Of all the papers in this special issue, Herbrand's shows most clearly men's paternal investments in children. Research on ARTs and parenting is usually carried out by women scholars, who work exclusively with women interlocutors. Herbrand's study of collaborative families is thus a refreshing departure. In general, we need to put men back into our reproductive and parenting studies. For example, some men are themselves infertile, and can only become fathers through resort to ARTs (Inhorn 2012). Similarly, other men who are fertile may end up investing heavily in ARTs to support their partners' paths to parenthood. Although half of the world's reproducers are men, their participation in the ART and parenting worlds has barely received the meaningful scholarly recognition that it deserves.

Finally, this special issue turns to the future of reproduction and parenting, as clearly laid out in the paper by Gunnarsson Payne and Erbenius. With the advent of oocyte cryopreservation (i.e. egg freezing), transmen can now become fathers using their own gametes, as can transwomen who have frozen their sperm. Transgender parenthood is no longer inconceivable. Thus, as Gunnarsson Payne and Erbenius' paper shows, health professionals are beginning to 'concretize' fertility care practices in anticipation of increasing numbers of trans patients.

Although the future of trans reproduction remains unclear, it is beginning to be discussed in scholarly and activist circles, as was clear at the recent conference on 'Remaking Reproduction: The Global Politics of Reproductive Technologies', held at Murray Edwards College, University of Cambridge, from 27 June 2018 to 29 June 2018. There, five scholars and activists participated in a 'Trans Roundtable', bringing to the fore the ethical, legal, and experiential issues surrounding transgender reproduction. Not only is fertility preservation through sperm and egg storage now possible, but the advent of uterine transplantation points to a day when transwomen may be able to bear their own children.

In conclusion, the beauty of this special issue is that it asks—and then answers—so many pressing questions about the contemporary state of technologically assisted procreation and what these ARTs mean for parents. Ultimately, this special issue should be required reading for all anthropologists and sociologists of reproduction as well as parenthood studies, plus the clinicians, policymakers and activists who hope to bring about a more just reproductive world.

References

De Geyter, C. 2018. "More than 8 Million Babies Born from IVF since the World's First in 1978." https://www.eshre.eu/ESHRE2018/Media/ESHRE-2018-Press-releases/De-Geyter

Faircloth, C. 2013. *Militant Lactivism? Attachment Parenting and Intensive Motherhood in the UK and France*. Oxford: Berghahn Books.

Faircloth, C., and Z. B. Gürtin. 2017a. "Fertile Connections? Thinking Across Assisted Reproductive Technologies and Parenting Culture Studies." *Sociology* 52 (5): 1.

Faircloth, C., and Z. B. Gürtin. 2017b. "Introduction – Making Parents: Reproductive Technologies and Parenting Culture across Borders." *Sociological Research Online* 22 (2): 1.

Gürtin, Z. B. 2014. "Assumed, Promised, Forbidden: Infertility, IVF and Fatherhood in Turkey." In *Globalized Fatherhood*, edited by M. C. Inhorn, W. Chavkin, and J.-A., Navarro, New York: Berghahn.

Gürtin, Z. B. 2016. "Patriarchal Pronatalism: Islam, Secularism and the Conjugal Confines of the Turkish IVF Boom." *Reproductive Biomedicine & Society Online* 2: 39–46.

Inhorn, M. C. 1994. *Quest for Conception: Gender, Infertility and Egyptian Medical Traditions*. Philadelphia, PA: University of Pennsylvania Press.

Inhorn, M. C. 1996. *Infertility and Patriarchy: The Cultural Politics of Gender and Family Life in Egypt*. Philadelphia, PA: University of Pennsylvania Press.

Inhorn, M. C. 2012. *The New Arab Man: Emergent Masculinities, Technologies, and Islam in the Middle East*. Princeton, NJ: Princeton University Press.

Inhorn, M. C. 2015. *Cosmopolitan Conceptions: IVF Sojourns in Global Dubai*. Durham, NC: Duke University Press.

Inhorn, M. C., and P. Patrizio. 2015. "Infertility Around the Globe: New Thinking on Gender, Reproductive Technologies and Global Movements in the 21st Century." *Human Reproduction Update* 21 (4): 411–26.

Spar, D. L. 2006. *The Baby Business: How Money, Science and Politics Drive the Commerce of Conception*. Boston, MA: Harvard Business School Press.

Thompson, C. 2002. "Fertile Ground: Feminists Theorize Infertility." In *Interpreting Infertility: Childlessness, Gender and Reproductive Technologies in Global Perspective*, edited by M. C. Inhorn and F. van Balen. Berkeley, CA: University of California Press.

Thompson, C. 2005. *Making Parents: The Ontological Choreography of Reproductive Technologies*. Cambridge, MA: MIT Press.

Index

Note: Page numbers in **bold** denote tables.